L. (Lucien) Baudens

On Military and Camp Hospitals

L. (Lucien) Baudens

On Military and Camp Hospitals

ISBN/EAN: 9783337161286

Printed in Europe, USA, Canada, Australia, Japan

Cover: Foto ©berggeist007 / pixelio.de

More available books at **www.hansebooks.com**

ON

MILITARY AND CAMP HOSPITALS,

AND THE

HEALTH OF TROOPS IN THE FIELD.

BEING

THE RESULTS OF A COMMISSION TO INSPECT THE SANITARY ARRANGE-
MENTS OF THE FRENCH ARMY, AND INCIDENTALLY OF
OTHER ARMIES IN THE CRIMEAN WAR.

BY

L. BAUDENS,

INSPECTOR AND MEMBER OF THE COUNCIL OF HEALTH OF THE FRENCH ARMIES,
FORMERLY SURGEON-IN-CHIEF, AND FIRST PROFESSOR, OF THE
PERFECTING SCHOOL OF VAL DE-GRACE, ETC., ETC.

Translated and Annotated by

FRANKLIN B. HOUGH, M.D.,

LATE AN INSPECTOR OF THE U. S. SANITARY COMMISSION.

NEW YORK:
BAILLIÈRE BROTHERS, 440 BROADWAY.

LONDON:	PARIS:
H. BAILLIERE, REGENT ST.	J. B. BAILLIERE ET FILS RUE HAUTEFEUILLE.
MELBOURNE:	MADRID:
F. BAILLIERE, COLLINS ST.	C. BAILLY-BAILLIERE, CALLE DEL PRINCIPE.

1862.

Entered, according to Act of Congress, in the year 1862, by
BAILLIERE BROTHERS,
In the Clerk's Office of the District Court of the United States for the Southern District of New York.

W. CRAIGHEAD,
Printer, Stereotyper, and Electrotyper,
Caxton Building,
81, 83, and 85 Centre Street

To the
ADMINISTRATIVE AND MEDICAL OFFICERS
OF THE
ARMY OF THE UNITED STATES,
THIS LITTLE VOLUME OF
Results of
SANITARY INSPECTION AND CONSERVATIVE SURGERY
IN THE
French Army of the Crimea,
IS RESPECTFULLY DEDICATED
BY THE
Translator.

TRANSLATOR'S PREFACE.

THE acknowledged value of the experiences of the French Army in the Crimean war, has suggested the translation of the present volume, with the desire to render the results of the dearly-bought lessons of that campaign useful to the American Armies in the present war for the preservation of our national Government.

It became the painful duty of the historians of the Crimean war to record many errors and oversights, resulting in a most fearful loss of human life. When these faults were discovered, their remedy was attempted by the medical and administrative officers of the army, with as much success as their resources allowed; and the expedients adopted for relief, in a wild and desolate region, at a great distance from their supplies, are at once suggestive and profitable to every person who may be concerned in the health of armies. In the translation, the French weights and measures have been mainly changed to their corresponding values known and used in the United States, precisely or approximately, according to the original intention of the author. In rough estimates the metre has been called the yard, an allowance being made when the numbers were large; but in every case where precise quantities are expressed, the rendering has been carefully made, and in some instances both denominations have been retained.

Money has occasionally been allowed to remain in francs, and decimals have been used in expressing parts of a given measure, as pounds, gallons, and feet, instead of the lower denominations.

BIOGRAPHICAL NOTICE.

JEAN-BAPTISTE LOUIS BAUDENS, the author of this volume, was born at Aive (Pas-de-Calais), in 1804, and received the degree of Doctor of Medicine in Paris, in 1837, having previously served in the medical corps of the French Army in Algeria. He was employed successively in the hospitals of Lille, Strasbourg, and Paris, and obtained in each the prizes in Surgery and Anatomy. As Aide-Major in Africa, he succeeded in forming a hospital of instruction at Algiers, where he gave clinical lectures. Upon his return to France, his eminent services in Africa, which had been witnessed by the princes of Orleans, gained him rapidly a fine position in the profession, and a wide circle of beneficent influence. He became Surgeon-in-Chief of the military hospital of Val-de-Grâce, in Paris, where he remained ten years.

On the 25th of July, 1855, he was appointed Medical Inspector to the French Army in Corsica, Italy, and the Crimea, and his subsequent history is detailed in this volume. The intelligence and devotion with which he engaged in this service gained him the approbation of his government, but cost him his life. Returning to France, he prepared the volume now given to the public in an English dress, but the miasmatic exposures which he had encountered in the camps and hospitals of Constantinople and the Crimea, planted the seeds of disease, which terminated in death from cancer of the liver on the 27th of January, 1858. He was the author of more than twenty works, chiefly upon military and hospital surgery.

AUTHOR'S PREFACE.

WHILE on my mission in the Army of the East, Marshal Vaillant, Minister of War, did me the honor to write:—

"We wish you to turn to account the important mission with which you have been intrusted in the East, and desire you to report what you have seen of the state of our military and field hospitals, and how our establishments for restoring health in the army of the East compare with those of previous wars, the efforts of the hospital service, and everything upon which our physicians have bestowed so much zealous devotion, intelligence, and heart. You will be expected to describe the diseases that prevail, and those that we should apprehend; the means employed to prevent or remove them, and the operations of surgery, with their successful or fatal issue.

"You will report your views concerning the present organization of the health service, and its operation in our hospitals, the army, and the interior, and upon the improvements that should be made in it. I shall attach great importance to your views upon these subjects. The English and Sardinian armies may furnish facts for comparison, if not for instruction, and upon this subject we should take large and liberal views. I will not presume to lay down a plan, but, my dear doctor, you see what I desire, and no Medical Inspector has for a long time been placed in so fine a position for rendering substantial services to Science."

On the same day I received from Baron Charles Dupin, Senator, and Member of the Institute, another letter, of which the following are extracts:—

"The Medical Service of our armies was never better organized, its agents better instructed or more zealous, and its materiel more complete, or in better state for use. A report from you, as Inspector General, upon the condition of the Medical Service in the Crimea and at Constantinople, its extraordinary services under fire, in three battles, and in a memorable siege, the assistance prepared at the theatre of combat, and in places where inventive genius and a love of humanity must supply all our wants, so far away from France, and in a rude season of the year. In addition to these, the development of epidemics,—the cholera and typhus, may double the ravages of war; and there will be afforded you an opportunity for rendering sublimer services than ever fell to the lot of Larrey, your predecessor, in the history of our armies.

"I have given you, sir, a brief and imperfect programme, drawn at too great a distance, and with a hand little competent, but with an earnest and vivid regard for the honor of France, and the glorious titles earned by the armies which you have the honor to represent. I doubt not but that in our army of the East, Science will attain new facts in the art of healing. You will assuredly collect and weave them into a picture, in which Art shall recognise the master's hand."

These noble suggestions pointed out to me an end so elevated that I dare not hope to attain it, but they decided me in undertaking this labor, which the *Revue des Deux Mondes* first published,* and which I here collect in a volume.

The grand memories of the Crimean war belong chiefly to history; and now wise and useful measures, as well as errors and faults, may be discussed with equal loyalty and double profit, to the end that henceforth, instructed by experience, we may be able with certainty to adopt the one, and carefully avoid the other.

* The numbers for February 15th, April 1st, and June 1st, 1857.

THE CRIMEAN WAR.

PART I.

THE CAMPS.

The war of the East, so full of instruction in military science, was not barren of medical teachings. Its field of observation, vast indeed, and often melancholy, furnished opportunities for submitting to decisive proof, and sometimes even for settling, grave problems of hygiene, medicine, and surgery, which had until then remained in doubt. The army profited by these new discoveries, which lightened the pains of the sick and wounded, and its ills were relieved or altogether removed by its surgeons, whose ceaseless devotion, and unwearied zeal, have many times deserved the warm encomiums bestowed by the Commander in Chief, and the Minister of War. Their labors resolved themselves into three classes, viz. prophylactic measures, or those employed to prevent diseases; the care of those wounded in war; and the treatment of diseases, which, as we know, made terrible ravages. This threefold duty points out three great centres of experience—the camps, the field hospitals, and the regular hospitals of the army.

On the 25th of July, 1855, I was appointed Inspector of the Health Service of Corsica, Italy, and the Army of the East. Having inspected Italy and Corsica, I departed for the Crimea towards the end of September, and thus found myself, after the taking of Sebastopol, upon the theatre of war. This war presents to the mind two impressions: the one glorious in brilliant feats of arms, the other mournful from passive sufferings. Of the one, the world knows the fullest details; while of the other, it has but vague ideas. Let us enlighten this subject, and we shall learn, that it is not alone in the assaults of battle, that our soldiers always display the greatest courage.

CHAPTER I.

MEDICAL TOPOGRAPHY OF THE CRIMEA.

THE steamer upon which I embarked, had on board a battalion of the 11th legion, and when we reached Malta, thirty men had already been attacked by cholera. They were placed in charge of the Sisters of Charity, and sent to the Lazaretto, where a small hospital had been arranged for soldiers taken severely ill on the passage, and unable to continue their voyage, without endangering themselves or their comrades. It would have been desirable to form a great hospital establishment at Malta, for the army of the East, but the resources were wanting, even to the English. Still, the few beds which the Lazaretto contained, were, as a sanitary measure, of great value; and at the end of the war, when typhus, imported from the Crimea, threatened to spread in the south of France, and through the fleets, we were able to escape the dangers of infection, by leaving at Malta a certain number of typhic patients. The governor of the island, as well as the consul of France, Mr. Henry Fourcade, left nothing undone. The governor not only received foreign soldiers upon the island, contrary to traditional usages and positive instruction, but even raised the quarantine for our benefit; nor did they have cause to repent of this, for neither cholera nor typhus gained a hold in Malta, although its hospital received many times, both choleric and typhic patients.

Malta is two days' sail from Smyrna, and the cholera continuing, four dead bodies were thrown into the sea.

Uneasiness and alarm began to prevail. The young soldiers who encumbered the decks, had no other covering but the sky, and their clothing was saturated with the dews of the chilly nights, succeeding the tropical heat of the days. I caused to be distributed warm wine at nine o'clock in the evening, and an infusion of tea at four in the morning. A salutary reaction checked the progress of the malady, and restored gaiety and songs. Some cholera patients, already cold and purple, were recalled to life by these stimulants, and twelve who remained, were placed in the small French Naval Hospital at Smyrna.

We found at Smyrna a magnificent barrack, sufficient to accommodate six thousand men, which the Marshal St. Arnaud had thought of changing into a hospital for our troops. A noted locality of thermal waters, called the *Baths of Agamemnon*, occurs a few miles from Smyrna, and the road thither leads past the site of an ancient temple of Esculapius, whose vast ruins indicate a gigantic edifice, now replaced by a Jews' cemetery, destitute of shade or monuments. The road is excellent, and affords easy communication between the city and the baths. The marshal's project would therefore have been excellent, had it been at once adopted; but he was delayed by groundless apprehensions of the salubrity of the country, and the English meanwhile, more resolute, established a very large hospital at Smyrna.* In the

* The English Civil Hospital at Smyrna, consisted of a large Turkish barrack, situated close to the sea, and almost on a level with it, at the south-west angle of the city. The building formed three sides of a square, open to the bay, so as to be freely exposed to the sea breeze and the summer wind, the "Imbat," which blew directly from the sea to the hospital from 9 A.M. to 6 P.M. The general sanitary condition of the locality was decidedly bad, and the wonder was that pestilence, in some form, was ever absent from the coast. By careful cleanliness, the hospital was kept measurably free from malarious influences. Sanitary regulations were ordered by Dr. Sutherland, March 28, 1855, requiring the streets adjacent to be cleansed daily, and nuisances to be abated; the street drains to be cleansed and covered; the drains close to the hospital to be cleaned by flushing, and, if necessary, to be relaid; the privies to be disinfected, and water-closets substituted; ventilation to be thoroughly extended, and all excreta to be thoroughly removed.—TR.

winter of 1856, having no more sick at that place, they stationed there a brigade of infantry.

In leaving Smyrna, we passed near Tenedos, not far from the tombs of Ajax, Hector, and Achilles, and within sight of Mount Ida, and the plain *where Troy was*. The name of every village and locality recalled a classic memory. We entered the Dardanelles, passed Abydos, on the Asiatic side, and four miles further the quiet of this region was replaced by great activity, the English being there constructing a hospital for three thousand sick. The site was well chosen to meet, at the same time, the demands of health, and capacity for defence. Further on we found two hospitals, one belonging to the English, intended for three hundred sick, and the other French, with four hundred and twenty beds. The latter had been established in a Turkish Lazaretto, and they had unfortunately built four large rooms in the middle of the central court, instead of placing them without, upon a little hill, which sloped gently towards the place. This error prevented the circulation of air, and was the more to be lamented when they were afterwards required to receive patients infected with Hospital Gangrene.

Upon landing at Gallipolis, I visited the French Hospital, and found another fault in the arrangement. To turn to use a considerable stretch of wall, they had begun by building sheds against it, at the foot of a hill, without noticing the drainage of the soil; but were shown their error, when cholera appeared on the premises. These blunders, unfortunately not rare, might have been avoided by consulting physicians, who, knowing the efficacy of air constantly renewed, especially in warm climates, where the rigors of winter are not felt, select elevated and not sheltered sites; but the quartermasters for barracks were not always occupied with these views, and, as in France, might be said to understand everything but hygiene. If the pupils of Saint Cyr* were to devote only a dozen hours to hearing a dozen

---

* Saint Cyr is the great military school specially devoted to the French cavalry service.—Tr.

lectures upon health, they would carry into the army certain principles of science, from which the soldiers would derive substantial benefits; the advice of physicians would be oftener heard; and the perils of epidemic diseases, to which our armies are constantly liable, would often be avoided. The organization of these hospitals was at least blameless; the furniture, beds, food, and fixtures left nothing to be desired; and the medical service was perfectly well directed. Everything bore evidence of an active zeal and an enlightened solicitude.

We crossed the Sea of Marmora during the night, and at the dawn of day, the splendid panorama of Constantinople, and of Seraglio Point, opened to our view. The slender shafts of the minarets, piercing the azure sky, were ranged like a guard of honor around the grand mosques which commanded St. Sophia; the air was filled with a thin vapor, and the landscape, dotted with pavilions and crowned with a cypress grove, seemed like a realization of the reveries of the *Thousand and One Nights*. It is a pity that the charm should vanish as soon as one steps into the labyrinth of its narrow muddy streets, full of quagmires, and overrun with street porters of Herculean strength (a race of biped camels that they call *hamals*), with dogs, and with asses laden with boards. The houses are of wood, of miserable aspect, and without architectural style or character.

I at once visited the hospitals, where the sick, who had come from the Crimea, were chiefly tainted with intestinal diseases, intermittent and remittent fevers, and especially with scurvy. In the wounded, the taint of scurvy impoverished the blood, which became more fluid, and oozed from the wounds in abundance, so that the most energetic appliances of science could not control these hæmorrhages, and they very often proved fatal. A still more terrible disease, the Hospital Gangrene, made fearful havoc. Many of the wounded from the Crimea, and those who had until now escaped, were tainted after a brief residence in the hospitals, and wounds almost closed, and considered as healed, were invaded by the gangrene. This scourge, which had but rarely been seen in Algeria, had, before I left for the east, already appeared in our hospitals at Marseilles and in the south, which received the wounded from the Crimea. It is contagious, and is transmitted by the air from one wounded person to another, so that in a ward impregnated with its miasm, it is very difficult to exterminate it.

On the 5th of October, 1855, I left Constantinople for the Crimea, on board the steam frigate *Descartes*, Captain Darricau, and left it with regret at Kamiesch. Reporting myself at the General's quarters and to Marshal Pelissier, I at once set to work, in studying the great questions of hygiene, necessary to be settled upon the approach of winter. I visited, upon horseback, the various positions occupied by our army, the camps, and the field hospitals, at the same time taking note of the medical topography of the country.

The part occupied by the Russians was almost entirely uncultivated, covered by immense plains, and without water; while that on which the Allies had encamped at the time of our arrival, was dotted with cultivated spots, and vineyards of considerable note. The soil was a dark vegetable mould, easily washed by the rains, and the mud of the Crimea defies all description. This layer of mould varies from one to several yards in thickness, and the subsoil is a porous limestone, easily torn up by the pick.* The regiments that were encamped in places where this subsoil appeared, dug circular pits about four or five feet deep, to place their tents, thus

---

* The geological features of the Crimea are concisely as follows:—
The western portion, extending as far east as a meridian passing through the head of Sebastopol harbor, consists of the Upper Tertiary or Steppe limestone, so often mentioned in this volume, and of a light porous structure. Along the shore, from Sebastopol to the western third of the southern coast, is a belt of volcanic sand and ashes; and from the monastery of San Georgeo westward, is a belt of erupted volcanic rocks, forming vast picturesque masses along the coast. The British and Sardinian camps, and the forests of Baïdar, were underlaid by Jurassic limestone; the borders of the Tchernaïa, from the Russian camps on the north, to the French camps and battlefields of Balaclava on the south, by chalk and green sand formations. A tract along the southern border from Col, eastward, is underlaid by schist and conglomerate.—Tr.

avoiding the cold, and especially the winds, which are continual, and sometimes terrible, in the Crimea.

On the 14th of November, 1854, the winds unloosed, and, attended with torrents of rain, threatened the allied armies with a great disaster. The tents, sheds, and shelters for provisions were overturned and swept level, in the camps, while the ships at anchor were beaten and shattered, and the *Henri IV.* and *Pluton* of our imperial marine, sustained irreparable damages in the Bay of Eupatoria. At Kamiesch and at Balaclava, the transports suffered great injury; and at the entrance of the latter port, eight large vessels, laden with munitions, provisions, and clothing for the English, were sunk, with part of their crews.

We ought not, however, to complain of the violence of the winds in the Crimea, for without them the soil would be constantly muddy, notwithstanding its surface drainage. The wind renewed the air of our camps, and swept off the mephitic gases with which the clothing was filled; it bore off the miasms of dead men and animals, buried by thousands, but which, although buried, could not decay harmlessly; and if the wind did not keep us from the typhus, it at least lessened its effect; and perhaps to it, we may have owed our exemption from the plague.

The district occupied by the Allies, measured five miles in breadth, by about fifteen in length. It was the ancient Tauric Chersonesus, of which Herodotus and Strabo have left descriptions, and like the Troad, it called up the memories of heroic ages. It was hither that Diana transported Iphigenia, snatched from the sacrificial fire, and here she made her a priestess; upon this inhospitable shore the tempests threw Orestes and Pylades, and where Iphigenia failed to sacrifice her brother. Henceforth other recollections will efface these ancient traditions, and the names of our victories will eclipse those of Orestes and his sister.

The roadstead of Sebastopol is striking from its extent, and its majestic appearance. It was, so to speak, staked out by the masts of sunken vessels, which reared their points as if to mark the place of a great ruin. Upon entering the roadstead, we meet the little Bay of Carenage, from the head of which the Russian vessels, during the siege, threw volleys of iron shot ; and two and a half miles beyond, is the mouth of the Tchernaïa, which flows among the reeds through a marshy country, and whose waters are in part diverted to feed the docks of Sebastopol. The fine road to Sympheropol, crossing the Tchernaïa upon a wooden bridge, winds up the hills, behind which was encamped the army of General Bosquet. It passes the spot where the English were encamped, when the Russians attacked them, and the battle of Inkermann was fought, November 5, 1854.

Prince Menchikof, whose army, elated by the presence of the grand-dukes Michael and Nicholas, had received large reinforcements, attacked the allied troops at three points simultaneously. While General Liprandi made a vigorous attack on Balaclava, the corps of General Dannenberg sought to gain the heights of Inkermann, taking in the rear the besieging army, which was being attacked from the other side by the battalions of General Timofeif that issued from Sebastopol, hoping thus to cut off our communications with Balaclava and Kamiesch.

We know of the heroic defence made by the English, as they were attacked about daybreak, in their camps, without any works of consequence to protect them. These valiant troops, commanded by the Duke of Cambridge, fought with valor the compact masses pressed against them. The corps of General Bosquet coming up in haste, turned the Russian columns by bayonet charges, while General De Lourmel vigorously repulsed the Russians under the walls of the city, and met a glorious death. General Canrobert, who planned by wise arrangements all the movements, was wounded in the elbow, but fortunately not dangerously. Our ambulances collected about 500 of the enemy's wounded, and it is estimated that 6,000 dead bodies of the Russians were left upon the battle-field. The English lost three generals, a hundred officers, and a multitude of soldiers; and our loss, though much less, included about 500 wounded. After this great battle, we cut up the road, here and there, with enormous trenches, to render another surprise more difficult. On their part, the Russians destroyed the bridge, to cover their retreat. At the foot of the mountain, from the top of which the Russian battalions were thrust by our soldiers, was a true ossuary, the bodies having been devoured by vultures, leaving the skeletons only of the men to await burial. These skeletons had belonged to men whose stature was not great, but whose frames were remarkably firm; and the femur or tibia of a Russian, very easily distinguished him from a Frenchman or an Englishman.

The valley of the Tchernaïa ascends towards the east, to the forest of Baïdar, with an average breadth of about four hundred and fifty yards. Its air is sickly, and it was impossible to improve it in the whole course of the war.*

From the tops of the hills which bordered the valley, the hostile camps were in view of each other, and after the taking of Sebastopol, some of the soldiers of the advanced posts established communications with one another, by means of white handkerchiefs fixed to the

* The valley of the Tchernaïa is thus described by Dr. Sutherland, in the Report of the Sanitary Commission dispatched by the British Government to the seat of war in the East.
"To the north of the ridge, forming the northern margin of the basin [of Balaclava], lies the valley of Tchernaïa. From the crest of the ridge the ground falls gradually to the foot of a chain of hills called 'Fedoukine Heights,' which rise rapidly to an elevation of 500 feet above the sea level. From their summits there is a rapid descent to the level of the Tchernaïa. The valley through which the river flows is broad and tolerably flat. The bottom is chiefly of marl mixed with pebbles and chalky debris, and the bed of the river, which is only a few yards wide, and is scooped out of the debris to a depth of four to six feet. Most of the ground is perfectly firm in ordinary states of the weather, but as the river approaches the head of Sebastopol harbor, the ground becomes a marsh. * * * The greatest breath of the valley, measured from its southern boundary ridge to the foot of Mackenzie's Heights, is about six miles, and the length of the wider part below Tchorgoun is about five and a quarter miles to Inkermann Castle, where the precipices of the plateau approach those of Inkermann, and between them lies the marsh at the mouth of the river. The whole of the valley is covered with grass and flowers, and there are no trees except in the marsh."—TR.

points of their bayonets. To these visits succeeded exchanges, the Russians having brandy, and the French bread and tobacco; they became very fraternal, and this kind of amicable warfare extended so far, that General Camou, of the 2d Corps, found it necessary to cool off the heat of these effusions by punishments.

On entering the valley, we noticed on the left, the Mackenzie's Heights, veritable walls, straight and inaccessible, with a central depression, as if to invite assault, but protected in the rear by three stages of commanding banks.* The Russians had thickly planted these escarpments with cannon, and from thence thundered the batteries named by our soldiers *Bilboquet* and *Gringalet*, which, after the battle of Traktir, made such sad havoc by firing upon our physicians and their servants, while busy in dressing and attending the wounded Russians. A similar occurrence happened after the battle of Inkermann, but the Russian government hastened to disapprove of these barbarous acts, and General Lüders did much to atone for them, in his correspondence with Generals Canrobert and Pelissier. These mistakes would be rendered impossible, if, by the common consent of nations, physicians and their hospital attendants were to bear a distinctive badge, alike in the armies of all countries, and such as would be easily recognised by both parties.

Further on we find the bridge of Traktir, by which the Russian columns sallied out the 16th of August, 1855. It is commanded by the Fedoukine Heights,† held by General MacMahon's corps. The valley is at this place some two or three hundred yards wide, and is traversed by two streams, the Tchernaïa, and a canal of supply, separated by a space of about a hundred yards. Its banks on either side were high, steep, and furrowed by ravines. In these positions, rendered still more

---

* These Heights form a long almost horizontal ridge of precipices of chalk, rising to the height of about 1000 feet above the level of the sea, and resting on a steep talus, extending from one end of the ridge to the other.—TR.

† These hills rise rapidly to an elevation of five hundred feet above the sea. They consist of loose chalk with layers of flint.—TR.

formidable by entrenchments covered with cannon, the French and Russian armies lay encamped face to face, and the disadvantages that would attend either of the two armies that should go to attack the other, were so great, that it would not be difficult to foresee the result of the combat of August 16th, while, moreover, the precautions taken by General Pelissier placed us beyond peril from surprise. At four o'clock in the morning, six divisions of Russian infantry, supported by 160 cannon, and three divisions of cavalry, under the command of Prince Gortschakof, passed the streams, upon several bridges, screened by fog, and attacked with fury our lines and advanced posts of Piedmontese. The conflict became general; the Russian columns, when turned back, were quickly rallied, and advanced repeatedly to the charge, but were again hurled back by the French bayonets, and crowded upon the bridges, across which they were beaten and driven by nine o'clock in the morning. General Pelissier reported 3,329 Russians killed, of whom 2,129 were buried by us, and there remained upon the field 1,669 wounded Russians, 38 of whom were officers, which were collected by our ambulances. Our loss was as follows:—19 officers killed, and 61 wounded; 172 non-commissioned officers and privates killed, 146 missing, and 1,163 wounded.

The whole of the left side of the valley was guarded by the French army, and even after the taking of Sebastopol, the divisions of the first corps, under General Salles, extended as far up as into the forest of Baïdar. A change of bivouac, and location in the woods, proved favorable to the health of the soldiers. The forest has a rich vegetation and frowning aspect—it has for its outline majestic mountains, furrowed by valleys, and picturesque hamlets dot the slopes. Here and there we notice elegant hunting lodges, such as the Chateau of Perouski; and the springs of fresh and clear water in this region form the sources of the Tchernaïa. No better bivouac could be found than in this forest, where the division of Autemarre remained in the best condition throughout the severe winter of 1856, although one night in three was passed on guard.

If from the forest of Baïdar we turn our steps to the east and south, we come by a series of mountains, where the Sardinian army was encamped with its field hospitals, to the little port of Balaclava, hidden in a sinuous opening between immense rocks.* It was formerly a secure retreat for pirates, and we may still trace the ruins of a fort built in the thirteenth century by the Genoese. This place contained only a few families of fishermen when the English arrived, bringing with them their industry. They built a railroad, and a hundred ships discharged continually the products of civilization. The camps of our allies were largely supplied with every article of need, by which they escaped the scurvy and typhus in 1856. When we compare the destitute condition of the English army at the opening of the war, with its state in 1856, we cannot but admit the vigor of the British government.

The general's quarters in the Sardinian army were fixed at the Greek village of Kadikeuï, which was invaded for the time being by a population of cosmopolite merchants; and General Marmora invited me to visit the field hospitals of the Piedmontese army, whose management and attendance deserved nothing but praise.

Between Kadikeuï and the valley of the Tchernaïa, extends an undulating field, where, on the 25th of October, 1854, was fought the battle of Balaclava. In the morning, the troops under General Liprandi, sustained by powerful artillery, and numerous squadrons, took four unfinished redoubts, which the Turks had aban-

* Dr. Sutherland thus describes the topography and geology of the port of Balaklava:—
"The coast line of mountain-ridge, is at this point broken across by a deep sigmoid fissure, forming the entrance of the harbor of Balaclava. On the west side of the entrance, the rock sinks almost perpendicularly into the water, and on the east side the entrance is bounded by a singular conical hill 469 feet high, on the summit and side of which, are the ruins of the old Genoese castle of Balaklava. The western slope of this hill consists of masses of the same compact reddish limestone, and the bulk of the hill itself is formed of detached fragments of the same rock, or rather of a brecciated form of it, resting on highly inclined beds of conglomerate."—*Report of Special Sanitary Commission*, p. 220.—Tr.

doned after a vigorous defence, and began a series of cavalry combats, in which the advantage remained with the English, when an ill-timed or badly-interpreted order saddened the events of the day. The Russians having taken some cannon near the redoubts, the light horse under General Cardigan received orders to re-capture them, a thing next to impossible. These splendid squadrons rushed forward, sabred the cannoniers of the Russians at their pieces, and swept through compact columns, but assailed on every side by overpowering numbers and a storm of shot, they turned upon their course, and opened again a passage through the ranks of the enemy, until, with the aid of the African Chasseurs of General Morris, they regained our lines with only half their numbers, the rest being left upon the battle-field. It is known, that the order to charge, emanating from Lord Raglan, was taken to General Lucan by Captain Nolan, who was killed in the charge, and it is altogether probable that it was not imperative.

In passing south, over the high plateaux which extend along the sea near Balaclava, we come to the Cape Fiorente of the Genoese, the ancient Cape Parthenion, where are found some traces of the temple of the Tauripolitan Diana, in which king Thoas sacrificed strangers. Near these ruins, in a recess among the rocks, and sheltered from the north wind, is the beautiful monastery of Saint George, the asylum of pensioners retired from the Russian fleet. Although a fine hospital might have been established here, the allied armies respected with care this monastery; its religious ceremonies were never molested, but were performed daily in the presence of a heterodox crowd, attracted thither by the beauty of the scenery, and the chime of its bells, which recalled the memories of distant lands. The elevation of the ground, with its gentle slopes, and the purity of the air, always freshened by the sea breeze, indicated this place as an excellent location for an encampment. Between Saint George and Kamiesch, the French cavalry were cantoned, and suffered much less from disease than the infantry, because the place of habitation was more healthy; the personal habits of the cavalier are more careful than those of the foot soldier; and they are more in open air, instead of being squat in their tents a large part of the day.

The plain of Kamiesch extends to Cape Chersonesus, and is bordered by two bays, called Kazak and Kamiesch, of whose existence our navy were at first but imperfectly informed, and whose discovery was truly fortunate. They were constantly filled with vessels coming with provisions for the French army, and upon these naked shores of easy access, enormous store-houses for provisions were erected, while the shanties of merchants, more or less respectable, were grouped each day more thickly around our military establishments. In a few months, the whole city had been as it were extemporised, with its broad and regular streets, its coffee-houses, its theatre, its police, its Catholic church, and even its Protestant temple. I have not spoken of its hospital, but this was well arranged and provided for, and its medical service was skilfully conducted. In it we found a great variety of diseases, because the men who were placed here, were found too ill for the time being, to leave for the hospitals at Constantinople.

Kamiesch is almost five miles from Sebastopol; and as we approach it, the ground is found thrown up by the works of approach, and strewn with great quantities of projectiles. These lay literally in piles, in the ravines of Carabelnaïa, and the little valley where the cemetery of Sebastopol is situated, and which was so often taken and retaken during the war. Its marble shafts were broken, its funereal urns dashed to pieces, its wooden crosses mutilated, and its tumulary monuments overturned: everything was devastated, but by cannon alone, for, notwithstanding the intensity of the cold, our soldiers respected the oaken crosses which stood over the graves.

Nothing is more affecting than to pass around a city taken after a long and bloody siege. Before Sebastopol, there were seen here and there, immense funnel-shaped pits, which had been made by the fire of mines, countermines, and camouflets, and everywhere we met trenches for ambuscade, where the French sharpshooters lay in wait from the dawn of day till night, watching the soldiers and officers, and the artillerists who appeared on the fortifications, or showed themselves before the cannon embrasures. At length the latter were closed by mats, made of ropes skilfully braided so as to be bullet-proof. At times, the French sharpshooters, only fifty yards from their enemies, would draw them into conversation. The Russians were provided with excellent half-boots, of which they knew our soldiers were very fond, and showing the point of their feet they would call out in good French, " *Venez les prendre.*"* It may be certain that our soldiers were not slow to reply, nor were they wanting with civil words. At another time, they would hold up over the battlements on one side or the other, a bottle or a flask, and the first who hit the object, was warmly applauded by all the marksmen. There was no hatred between the Russians and the French; if they did not kill, they embraced.

In entering Sebastopol by the Mat bastion, I paid a silent homage to the skill of General Todtleben. Before me arose the tower of Malakof, and the Mamelon-Vert, taken on the 7th of June, 1855. The formidable ramparts of the latter, were raised before the former, and arrested our labors of approach, until after a bloody struggle of some hours, when having been taken and retaken several times, it remained finally in our possession. As the French assailed the White works, the English gloriously took the position called the Quarries; but notwithstanding the skilful arrangements of General Pelissier, we approached Sebastopol but about four hundred yards. The day was unhappily saddened by the imprudent ardor of the regiments, who ran up to the Malakof, and whose heroic courage was punished by many and much regretted losses.

I visited the fortifications of Malakof, a true labyrinth of Italic S's formed in relief by the earth brought up from the subterranean blinds, and the arti-

* Come and take them.

ficial mounds rising each day, at length exceeded the height of the tower itself. During the last days that preceded the assault, the besiegers hurled against the bastion 6,000 bombs every twenty-four hours. The Russians were unable to come out; in burying a corpse, they sacrificed two lives, the dead were therefore oft without sepulture. Meanwhile their casemates sustained by enormous masts taken from the fleet, and covered with many yards of earth, could not be injured by this terrible hail of projectiles; the bastion stood out like an aerial island in front of the Mamelon-Vert. On the 8th of September, our soldiers entered it by an irresistible rush. Nothing could stop them; neither the broad ditches surrounding it, nor the high parapets, bristling with bayonets and cannon, nor the firm heroism of the defenders. Without waiting for ladders, they sprang into the ditches, and climbed the parapets, by mounting upon the shoulders of their comrades. These positions once gained, they for five hours defended them against the vigorous efforts of the Russian masses, who struggled to regain their lost ground. In the absence of General Todtleben, detained by a wound, at a distance from the defensive works, the Russians had committed the very serious mistake of closing the gorge of the Malakof, on the side towards the city, and of leaving only a narrow passage, through which their columns could not deploy, and these thrust themselves fruitlessly upon the bayonets of our soldiers, now become steady in proportion as they had just before been ardent and enthusiastic. The enemy had depended upon an electric wire, leading to some seventy tons of powder, to bury in the ruins of the bastion the besieging army, but by a happy accident the blow of a pick exposed the wire, and it was cut. A few moments after, it demolished forts Paul and Alexander, to cover the retreat of the Russians. Upon this day, which decided the war, we had 5 generals killed, 4 wounded, and 6 missing; 24 superior officers killed; 20 wounded, and 2 missing; 116 subaltern officers killed, 224 wounded, and 8 missing; and 1,489 non-commissioned officers and soldiers killed, 4259 wounded, and 1,400 missing. Total loss, 7,551.

The district occupied by the allied armies was about sixteen leagues in circuit, the surface mostly uneven, and cut in various places by deep ravines, affording water of good quality, and presenting everywhere places favorable for camp and military operations. Upon the sides of the mountains, the tents of the allied armies were traced in rising and picturesque outlines, and fine macadamized roads, built and kept up by our troops, afforded facilities for transporting provisions and ammunition. The sheds of the merchants formed small villages, which the soldiers, in recognition of the probity of these industriels, named Pickpocketville, Rogueville, &c. A thoroughly organized police inspected the wines and brandies, and we rarely heard complaints of their being adulterated.

The climate of the Crimea, excepting certain marshy spots that might be easily made healthy, is remarkably salubrious, and excepting the cantonments, which the necessities of defence exposed to the paludal influences of the Tchernaïa, our troops were in good hygienic condition. The heat of summer, tempered by the sea breeze, scarcely exceeded that of the south of France. The winters are severe, the thermometer sinking —20° Centigrade (—5° Fahr.), and lower, while the violence of the wind renders the cold very hard to endure.

Our armies found but few resources in the country, and the Tartars could only sell some cattle, sheep, chickens, and eggs. They were as greedy in a bargain as the merchants we have mentioned, and I have seen them ask five francs for a hundred walnuts—and they found customers too. We carefully respected their villages, disturbed nothing, and maintained good relations with them. The wood which covered the hills quickly disappeared, and in the winter of 1856 the roots of the stumps themselves were dug up. We have read of the passage of locusts in eastern countries, but the presence of an army is a scourge even more sweeping; for the former devour only what is *upon* the surface of the earth. The distance of over twelve miles, prevented us from turning to profit the rich forest of Baïdar, and the administration found it easier and quicker to bring its wood from Varna. After the taking of Sebastopol, the wood employed in the defences was largely used in the kitchens of the regiments, while the troops adjacent to Baïdar, alone, continued to obtain its wood from the forest, by the aid of *arabas* and oxen, which the authorities placed at their disposal. These arabas are the wagons of the country, clumsy and entirely of wood, not a particle of iron being used about them. They encumbered the road to Voronzof, and the creaking of their wheels was music to the ears of the Tartars. The Russians brought in their provisions upon arabas, drawn by oxen, and upon their arrival, the cattle were slaughtered, and the wood of the chariots served to cook them.

About six miles from Kamiesch, upon a hill, in the midst of the French army, were the general headquarters, and there was the modest little tent, in which General Canrobert passed the winter of 1855, having given up, for the use of the hospitals, the temporary barrack destined for the use of the Commander-in-chief. He wished to share with his men the rigors of the season. This example of self-denial in a high station, infused a heroic spirit through every rank; but not by this alone, but by many other honorable and ingenious devices, did the General-in-chief seek to sustain the spirit of the army, through the rudest vicissitudes of their duties. Upon the highest point of the General's quarters, there was built a clock tower, of stone, and the clock, with its dials, taken from a belfry in Sebastopol, served to regulate the time of their watches. Around the quarters of Marshal Pelissier were grouped the offices of the postal service, the treasury, the electric telegraph, the chief chaplain, the little extemporized church, where mass was said, and in short, all the chief departments of the service.

I soon entered into the details of the camps, regiments, regimental infirmaries, and the field and regular hospitals, and consulted the generals, supervisors, chiefs of departments, and physicians, to learn the wants of the soldier, and fix my ideas concerning measures relative to rations, shelter, and clothing.

On the 20th of October, I addressed to the Minister of War the following report:—

"To His EXCELLENCY THE MARSHAL.—I have devoted the first days after my arrival in the Crimea, to visiting the field hospitals of Divisions, to examining the camps, and to the study collectively of the great questions of hygiene, implied in the very important mission which your Excellency has deigned to intrust to me.

"I am now able to express an opinion upon some of these points, and I can with pleasure affirm, that the health of our troops was never better. In the Crimea, as at Constantinople, half of the beds in the hospitals are unoccupied, while the transmission of the sick has nearly ceased. We have neither cholera nor typhic fever—dysenteries are rare, and the scurvy is less intense. The hospital gangrene has begun to disappear, wounds assume a better aspect, and every day numbers of the wounded leave the field hospitals, to return to their regiments, or to their families, to await a formal discharge from the service.

"The beauty and salubrity of climate, the taking of Sebastopol, the elevated tone of military spirit, the constant watchfulness of the Commander, the wise measures adopted by the military intendants, and the skill and devotion of the medical corps, are praised by all, and have led to these happy results.

"The field hospitals of Divisions are well adapted to their purpose, and each of them will accommodate 750 sick, sheltered by temporary barracks or tents. The rations of food, as a general thing, leave nothing to be desired; and if the bread is sometimes too damp, or a little burnt, it is explained by the difficulty of dealing out fresh bread every day, in a country where everything has to be brought, even to the wood; while as an offset, the wine is of superior quality, and is served at the table of the soldier, the same as to that of the general. The preserved vegetables have done us much service, and milk concentrated by the method of Lignac has been tried with satisfactory results. This mode of supply is the only one possible at this time, and would be of great use in our hospitals in the Crimea, and even at Constantinople, where milk is scarce, of very poor quality, and of high price.

"If the whole army should winter in the Crimea, there would be an absolute necessity of changing without delay, the field hospitals of divisions into temporary hospitals. It would be sufficient to increase the number of shanties, to abandon the tents, so precious in summer, and to furnish the shanties with furniture, bedding, and utensils of every kind necessary. Hospitals for 6000 sick, together with the regimental infirmaries, would supply the wants of the army; and the sending of patients to Constantinople, which until now has been so frequent, often so unpleasant to the sick, especially in rough weather, and always so burdensome to the treasury, would be necessary only in exceptional cases. The hospitals at Constantinople, which so greatly need to be purified by rest, would thus become a reserved resource for secondary use, and a part of their attendants and furniture might be sent to the Crimea. It would be wise to construct near the monastery of Saint George or at Constantinople, an extensive depôt for convalescents; for it is of first importance to put an end to these emigrant fleets. The best way to restore a moral tone, and put an end to longings for the domestic fireside, natural enough to the men, but enervating to the army, is to cut short this idea of sending off the sick, which I believe has been greatly abused, since, out of every hundred sick taken to Marseilles, only ten had need to enter the hospital. We might with little expense construct here ample barracks for the sick. The walls might be made of biscuit boxes filled with earth, and laid one upon another, or by the use of casks, or the fascines used in the siege, which might be laid up in a few hours, and covered with a roof of boards. This system of protection against the rain, snow and cold of winter, might be extended to the whole army; there would be no lack of fascines, and if boards are wanting, they can use their tents for the roofs.

"The imperious exigencies of the long and glorious siege of Sebastopol have happily ended, and though we still hear the thunder of cannon, the bullets are almost harmless, and the time has come when we should be actively employed with measures for sheltering from the rigors of winter, the soldiers, to whose heroism France, which honors every kind of merit, will hereafter, without doubt, raise one day an imperishable monument."

Pursuing my inquiry, I resolved to inform myself correctly on all points, to seek the means for warding off the dangers which the approaching winter threatened us with, and to ascertain the truth, from the varied and often contradictory opinions which were expressed upon these important questions.

CHAPTER II.

RATIONS.

IN my investigations concerning the provisioning of the army of the East, four principal classes were indicated by the nature of the alimentary substances themselves, and I have examined successively, the facts concerning bread, meats, vegetables, and drinks.

Biscuit, which is the bread of mariners, was of great reliance in our Crimean army. It is easily kept, if free from moisture, is easily carried, and weight for weight is much more nourishing than army bread. The water which bread contains, increases its weight a third part, while none of this remains in biscuit. Flour, when made into biscuit, is lightened five per cent. The biscuits from France were good, while those from Constantinople were not always of faultless quality, and sometimes contained traces of moisture.

The usual ration of biscuit, was 550 grammes (1.21 pounds), besides 185 grammes (0.4 pounds) in the soup.

At the opening of the campaign, the ration was increased to 650 grammes (1.43 pounds), and when the troops labored in the trenches, they received daily per man an increase of half a franc in pay, and of 250 grammes (0.55 pounds) of biscuit. Still, the soldiers preferred army bread, though heavy and coarse, to the best of biscuit, as the former digests more slowly, lies better on the stomach, and never leads to satiety and disgust. Biscuit, without leaven, is extremely absorbent, acting in the stomach like a sponge, and after having exhausted the salivary glands in mastication, it absorbs the gastric juice, so that not enough remains for proper digestion. For the purpose of softening it, they soak it for a moment in water, and then hold it to the fire, by which it becomes pasty, insipid, and indigestible. Biscuit should therefore not be distributed, unless bread is wanting. Very often half rations of bread are given, and during the Crimean campaign four distributions in seven, were of biscuit. It is no easy task to furnish fresh bread to an army of 140,000 men, when it is necessary to bring from beyond seas, the flour, the wood, the kneading troughs, and the ovens.

In Paris, the baker's civil tax is based upon the proportion of 130 pounds of bread to 100 pounds of flour, this ratio between the quantity of flour and water which it should absorb, being recognised as necessary for the proper preservation of the bread.* The war department imposes no limit, and the highest ratio is the best.

* In the American army, the regulations allow, under ordinary circumstances, the issue of flour at the rate of 22 ounces daily to each man, in place of the same weight of bread. This flour, drawn by the Division, Brigade, Regiment, or Post, is baked by soldiers detailed for the purpose, in ovens made of sheet iron, covered with brick, stone, or earth, or in ovens of brick alone, and the savings that result go towards establishing a fund for the benefit of the posts or regiments concerned. There are probably no regiments in the service, in which a sufficient number of masons and bakers could not be found, to construct and operate these bakeries. For purposes of economy, they are often worked day and night, by relays of men. The savings have in some cases exceeded 33 per cent., and will, in most if not all, come up to 30 per cent. The economy of the arrangement appears in the difference of transportation, and in the quality of the bread, as well as in the savings upon weight, and the facility with which negligence and abuses can be corrected, is so great, that they may be checked upon first appearance. In Gen. Kearney's Brigade, of four regiments in Franklin's Division of the army of the Potomac, the savings in two months, in the winter of 1861-2, amounted to $3,436.—Tr.

With tender wheat, 144 pounds, and with hard wheat, 150 pounds to 100 of flour, can be attained; but bread too much charged with water, bakes badly, the crust burns and blackens, it soon softens, loses its fermentation, and yields a food more or less defective. Perhaps bread made a half or a quarter biscuit, with a little leaven, should be prepared for armies in the field. The ration would be less in weight, but not less nutritious, and it would obviate, to a considerable degree, the inconveniences of biscuit.

The bolting of flour for army bread, has for some time been carried as far as 20 per cent. of bran extraction, for tender wheat. It was thought the less bran there was in bread, the more nutritious it would be, and that the whiter it could be made, the more it could be used for soups, instead of bread bought from the civil bakeries, in which the bolting is carried to 40 per cent. of bran. This innovation does not seem to be wise, for it raises the cost without a corresponding advantage. The new kind of bread digests too rapidly, and does not mix well in soups. It is not shown that bran, within certain limits, does not contain elements capable of assimilation, and in times of scarcity or war, it is a thing of no small consequence, to carry the bolting of bran to such an extent for an army of 500,000 men. Besides, our soldiers are chiefly from the rural districts, and would prefer bread of inferior quality, such as they have been accustomed to, rather than the whiter bread of our army supplies. Our Russian prisoners, accustomed to a very coarse bread, were not sufficiently nourished by the bread furnished to our soldiers, and required increased rations.

The best fresh meat is beef. Alone it makes a good soup, and according to a saying as true as it is common, *soup makes the soldier.* Our cattle arrived in the Crimea after such long vicissitudes, that they might be said to be as lean as Pharaoh's kine, and in order that the quantity might make up for the quality, the ration was increased from 250 to 300 grammes (0.55 to 0.66 pounds), but the bones, which were included, were of great relative weight. I recommended that the hard parts, after having been boiled in the pot, should be pounded up and boiled again, to extract the gelatine; and this method, as employed in the hospitals at Constantinople, very sensibly improved the broths of the sick, and we were able to order it as a regular prescription to the regimental and hospital cooks. In France, it is true, bones are sold, but here, would they be worth what it would cost to take care of them?

When fresh meat was wanting, it was replaced by preserved dressed meat, put up in tin cans hermetically sealed, and as it contained no bones, the ration was reduced to 120 grammes (0.25 pounds). These preserved meats were of excellent quality, but the soldiers disliked the change, for they estimated the weight and bulk more than the quality, and although the quarter of a pound fed them really as much, they declared it insufficient, and preferred fresh meat, even of poor quality. Sometimes the ration was sausage and lard, and occasionally packages of powdered meat. The latter was eaten by few, was liable to adulteration, had a suspicious odor, and the men were always in doubt whether animals of all kinds were not used in its manufacture. After a few days' use, the troops manifested a great repugnance and disgust towards it.

The sheep, which contrived to find a few blades of grass, although insufficient for the sustenance of cattle, were kept in good condition, and were highly prized. A great number of horses died during the winter of 1855–6, and following the example of the distinguished savan, M. Isidore Geoffroy Saint Hilaire, I preached the doctrine of eating horse flesh, but made few converts. The horse is herbivorous, like the ox; no animal is more neat, it is washed and cleaned daily; its flesh, although more firm, is not less satisfying, and it would make excellent soups. In Germany, horse flesh, cut up, is sold publicly in the butchers' stalls. The two artillery batteries of Autemarre's division, encamped at Baïdar, fed upon horse flesh; nor did they have reason to regret it, as they escaped the diseases and mortality which swept so cruelly through the rest of the army. Experiments made by very competent savans have proved, that horse flesh, even tainted, as in the black-tongue, when purified by fire, may be eaten without danger, but still, I would not venture to advise the eating of diseased horses. I know that the meat of oxen, flaccid, colorless, and gluey, which we were sometimes forced to distribute in times of scarcity, has caused diarrhœas.*

Fish, and especially the buckle turbot, are abundant on the coasts of the Crimea; and while butchers' meat, even of poor quality, sold at Kamiesch at a quarter of a dollar a pound, a turbot of ten pounds cost only seventy-five cents or a dollar. After the taking of Sebastopol, the officers, with nets found in the city, made wonderful hauls of fish in the bay of Streteska. I regret that we were unable to establish large fisheries, to contribute by this precious resource to the support of our army, and afford variety to its food. We also found excellent game, quails and woodcock, at the time of their migration, and hares, pheasants, and roebucks in the forest of Baïdar. The feast of St. Hubert was held in the forest. It is needless to say, that the luxuries did not go to the *soldiers'* tables. Some of the officers kept fowls for their eggs.

The want of fresh vegetables for our army was a great privation. Preserved vegetables were never wanting; and the mixed kinds, which we distributed very regularly, were the most relished of all. At the end of the campaign, these preserves were of poor quality, and were found sometimes so altered by fermentation that the soldiers threw them away. The avidity of the dealers was not checked by the miseries of the army, and tended greatly to increase them. The sacks of potatoes received from time to time, were very acceptable. The government delivered them at the rate of three cents the pound, but in the shops at Kamiesch they were sold

---

* Under favorable circumstances, cattle are issued to the American army upon hoof, their weight being determined by the following rule: From the live weight of a steer, there is deducted 45 per cent., when the gross weight exceeds 1300 pounds, and 50 per cent. when it is less than that, and not under 800 pounds. When pasturage or forage are abundant, the post or regiment may often realize a valuable saving, both in quality and quantity, by allowing their cattle to recruit from the exhaustion of a long journey by railroad and to gain flesh by feeding.—TR.

at from nine to twenty-five cents. For cabbages, as much as $1.86 were paid. Man has need of vegetables as much as of meat. Physiologists divide alimentary substances into two classes; nitrogenous, which, according to Dumas, supply the demands of assimilation, and non-nitrogenous, which furnish the products consumed by respiration, and are called by Liebig respiratories. The want of vegetables, which belong chiefly to the latter class, impairs the respiratory function, and injures by hematosis. It has been demonstrated, that the ultimate effect of this regimen would be death.

The necessity of a varied nourishment has been established as a practical fact. The theoretical views by which it has been attempted to account for this necessity, do not seem to accord with the facts observed in late years. The need of a varied food is recognised; explanations of the effects of the food, alone, ought to be abandoned. They are based upon very attractive views, which, recognising in vegetables the power of supplying the chemical elements, as oxygen, hydrogen, and nitrogen, or the simple compounds, as water and carbonic acid, and of making from these elements, what organic chemists call the proximate principles, as amidon, sugar, gluten, etc., deny to the animal, the power of fabricating these proximate principles. They seem to think, that animals should become the borrowers of the food ready prepared, and that they should limit their rôle to its assimilation. It is now proven by the beautiful experiments of M. Claude Bernard, on the formation of sugar by the liver, in animals fed exclusively upon meat, that they *can* manufacture proximate principles, as well as vegetables can. Furthermore, in showing, that instead of the blood's warming in the lungs, it actually becomes cooler, M. Bernard has rendered inadmissible the hypothesis, which would make this organ the seat of combustion, arising from the combustion of atmospheric oxygen with the carbon of venous blood.

Desiccated vegetables, having lost their water of vegetation, and perhaps other gaseous elements, which analysis has not been able to discover, cannot adequately replace the fresh articles.* In the army of the East, the imperfect aeration of the blood could be traced in sanguineous effusions, and in scurvy. As to food and habitation, the Crimean expedition might perhaps be compared to a long voyage; the army being, as it were, limited to a great ship, and subjected to the influences of navigation on a vast scale. Prolonged habitation in common, generated noxious exhalations, and finally typhus. The invasion of scurvy was delayed by the presence of a plant, as precious as it was abundant, the *Terrassacum* of Linnæus, commonly known as the dandelion. As the trees and their roots were removed from the soil of the Crimea, this plant became the king of vegetation. The Crimea was the Promised Land of the Terrassacum. It resisted bravely the destructive war which our soldiers waged upon it; torn up without stint, it sprang up again in still greater profusion. It made a salad of easy digestion, and with a pleasant and salutary bitterness. It appeared daily on the table of Marshal Pélissier, with whom it was a great dainty. Unfortunately, in the heart of winter, and middle of summer, the extremes of cold and heat checked the growth of this useful plant, and its disappearance was followed by the development of scurvy.† The Minister of War caused to be purchased in the markets of Constantinople, large quantities of fresh vegetables. One of my reports says, " 100,000 francs, spent in fresh vegetables, is a saving of 500,000 francs in the expense of sending the sick to

* The desiccated vegetables furnished to the American army, have in many cases fallen into disrepute, from the faulty manner in which they have been cooked. The error has generally consisted in omitting to soak them thoroughly beforehand, in not boiling them for a sufficient length of time, and in putting them into soups in too great quantities. When thoroughly soaked, sufficiently boiled, and properly seasoned, they are generally relished. The American army in Utah owed their preservation from scurvy largely to the use of these desiccated vegetables, at a time when nothing else of an antiscorbutic kind could be obtained.—Tr.

† Among the antiscorbutics used with success by the army of the United States, in western expeditions, have been agave juice, wild artichokes, lamb-lettuce, wild onions, the tender shoots of the Phytolacca decandra or pokeweed, nettles, celery, and wild oranges. Raw potatoes, sliced in vinegar, are highly useful.—Tr.

hospitals." Towards the close of our sojourn, we made kitchen gardens, and raised some few things of first necessity. They promise great resources in the future, should we remain in a country so destitute of everything. I am surprised that the army was not furnished with sour krout, which is so easy to keep. Lentils were scarce, but kidney beans plenty.

Vegetable acids, that is apples, lemons, and oranges, were wanting. These, as we know, are antiscorbutic. The English drew rations of lemon juice preserved in casks, and they used it in their grog, with rum and sugar. Our field hospitals and regimental infirmaries were well provided with it towards the close of the campaign, but our experience, although attended with good results, was not sufficiently tried to render it fully conclusive. The English physicians ascribe great antiscorbutic virtues to the lemon juice, and say it is in a large degree owing to its use that their army was saved from the scurvy, in the winter of 1856. It has for a long time been valued as an antiscorbutic by their mariners, who take it with them on long voyages.

Soup is excellent food for the soldier, but its quality depends much upon the cook.* Every soldier takes his turn in cooking, as he does in mounting guard; but this is wrong. In the same regiment, some companies eat good soup, and some bad. The army officers do not usually trouble themselves with these details, so important—for the first condition of health is that the stomach be satisfied. In the Crimea, the troops that best sustained privations and fatigues, were those commanded by colonels careful of their men. Let us take an example of two regiments, which left the camp of

* In the American service, the several companies of a regiment are provided with kettles and fixtures for cooking by companies, a cook and one or more assistants being detailed or hired for each company. Probably this service is never better performed, than where an experienced cook is steadily employed, and the assistants are changed at short intervals. By this arrangement, nearly every man in the company will have an opportunity of acquiring that culinary experience which would enable him, if detached from the company, and thrown upon his own resources, to prepare his food in a manner consistent with the preservation of health.—Tr.

Saint Omer at the same time, arrived together in the Crimea, in October, 1855, encamped side by side, endured the same vicissitudes of weather, and performed the same services; the one preserved on the 1st of April, 1856, 2,224 soldiers of an effective force of 2,676 men, while the other, with an effective force of 2,327 men, had 1,239 only left—and these losses did not include those wounded in the war! In the armed naval service, our commanders of vessels supervise the preparation of meals for the crews, and observe most punctually the hour for breakfast and dinner, which is never delayed, anticipated, or interrupted. It is earnestly to be wished, that the same scrupulous care might find its way into our land armies, and that these wise hygienic measures should never be infringed upon, except in cases of absolute and manifest necessity. We pay rewards to those colonels of cavalry whose squadrons preserve the greatest number of horses, and these rewards excite an excellent and profitable emulation. We could have similar, but more important and happy results, were we to grant similar tokens to colonels, whose battalions preserved the most men in the best health.

Wine does not form a part of the ordinary rations of the soldier in a campaign. That which was distributed to the army of the East was generally good. Each soldier drew half a pint, and officers were allowed to take daily from the stores, besides their rations, a litre (0.264 gallon) of wine, for which he paid fifteen cents, while the private dealers sold wine for three or four times that sum. During the epidemic, Marshal Pélissier doubled the rations. For our sick, we had generous wines, which the administration gave liberally. Brandy alternated with the wine, the ration being about a third of a pint. Taken immoderately, brandy is very dangerous in winter, and exposes the drunkard to perish from cold; but in moderation it excites a salutary reaction. M. Laurent, a ship lieutenant, with some marines, had charge day and night of a battery before Sebastopol; he preserved the health of his cannoneers through the winter by giving them, at stated intervals during the night, three warm grogs of brandy, through which the system gained great power to resist the cold. Coffee was often given out instead of wine or brandy, the rations consisting of sixteen grammes (about half an ounce) of coffee, and twenty-one grammes (three-fourths of an ounce) of sugar. During the early campaign in Algeria, the columns about to be sent out received advance rations of brandy for eight days, which were consumed before starting. The drunkenness which ensued was a lamentable prelude of the fatigues and privations of the war; and at the time of the expedition to Mascara, in 1834, it was necessary to send to the field hospitals a host of soldiers attacked with dysentery. Upon setting out upon the expedition of Tlemcen, I advised the substitution of coffee for brandy, and with very decisive results. Coffee has become among our soldiers in campaign a healthful and favorite drink, and is found to prevent the intestinal looseness so frequent in warm climates. The Arabs take daily, several light infusions of coffee, and when in their country, we ought to be governed by their traditional usages, founded upon reason. The soldier, by steeping some pieces of biscuit in his coffee, makes at will a very nutritious soup, of which he never gets tired. Coffee is especially useful on a halt, or in the trenches, and, in short, anywhere when the soldier has not time to prepare his soup. It refreshes and enlivens, while it does not prevent sleep after a day of fatigue in the open air. It recommends itself to the government, being easily kept and carried. It should not be ground long before use, because it then loses its volatile aromatic principles. It may be roasted and distributed in the grain. In the Crimea, we gave the troops little mills, which readily prepared it for infusion. The ingenuity of the soldiers furnished means often original, and not always prosaic, for preparing it. I have seen in our camps, the coffee ground by a ball rolled about in the half of a bomb shell. The English replaced coffee by tea, which their troops took morning and evening, seasoned with rum. Pieces of bread soaked in this grog afforded a nourishing and agreeable food. Thus we meet in the bivouac with the home usages of British families.

The following table shows the rations of an English soldier in the Crimea and upon the Bosphorus:—

In the Crimea.

Bread,	1½ pounds.	or cocoa,	1 ounce.
or biscuit,	1 "	or tea,	¼ "
Fresh meat,	1¼ "	Wood,	4½ pounds.
or salt meat,	1 "	or mineral coal,	1½ "
Sugar,	2 ounces.	or charcoal,	1 "
Rice,	2 "	Candles, per man,	2 ounces.
Coffee,	1 "		

On the Bosphorus.

Bread,	1½ pounds.	Mixed vegetables,	4 ounces.
or biscuit,	1 "	or potatoes,	8 "
Fresh or salt meat,	1 "	Rum,	½ gill.
Sugar,	2 ounces.	Candles, per doz. men,	3 ounces.
Coffee,	1 "		

On days when they issue salt meats, they give besides, two-thirds of a pint of peas or kidney beans, or a quarter of a pound of mixed vegetables, or half a pound of potatoes.

As an antiscorbutic, they give thrice a week to the troops, some lemon juice, and four pounds of sugar to every hundred men.

The French soldier was never for a day without food, the distributions were made as regularly as in a city garrison, and with as great variety as was possible in a country without resources, and eight hundred leagues from France, from whence everything had to be sent.

I will here offer some brief observations upon the improvements which appear to me practicable to introduce in the regimen of the soldier. Without exceeding the strict limits of expense, I believe that the number of sick may be diminished, and along with this the expense of hospitals, by giving a third meal, by varying the food, and by increasing its amount. Our soldiers take two meals: one at ten o'clock in the morning, and the other at four in the afternoon. The breakfast is eighteen hours after supper. Now a man may, if occupied with intellectual labor, take food perhaps only twice a day; but the young soldier, who has not yet reached his full growth, and who is exposed to a great expenditure of physical strength, ought to eat oftener. Before entering the service, he was a peasant or a mechanic, and, practising the popular adage, " never go to work hungry," ate before beginning his labors. When he joined the regiment, this morning repast, which he had taken from infancy, and which to him is a real want, is at once, and without gradation, cut off. The old soldier, whose stomach is less imperious, takes good care to eat a piece of bread and drink a little glass of brandy before going to his drill. The conscript may do so likewise; but always in a hurry, and compelled to regulate his life by the tap of the drum, will he always find time to take this irregular repast? If it is useful, why is it not made regular, and a time given to it in the distribution of the daily duties, as, for example, at seven in the morning? A sup of coffee, a piece of cheese, an onion, a little bread and butter, and some wine, would be sufficient.

When a man's food is not varied, his health quickly fails, as Magendie has demonstrated. The mariners who in a long voyage are reduced to biscuit and salted meat, readily contract scurvy, typhoid fever, and sometimes typhus. The soldier eats invariably twice a day soup, made of boiled beef, and vegetables, whose quantity varies with the price. To break this monotony, the men often sell their bread to buy fruits, or cheese; the ration of bread is, however, estimated from the wants of the system, and the sale of a part weakens the body, and procures a variety, without adequate nourishment. These two eternal soups are the strongest reasons, and I know this positively, why the soldier will not willingly re-enlist.

In 1847, a scarcity of provisions doubled the number of the sick; a fifth part of the effective regimental forces were in the hospitals and infirmaries, ninety-two scorbutic patients were entered at Val-de-Grace, and although the furloughs of convalescence were immensely numerous, the number of deaths per one thousand arose from fourteen to twenty-nine. During the same year, the crack regiment, the municipal guards, and the firemen, who were able on account of their supplementary pay to add to their rations, escaped the diseases which raged among the troops of the line, which depended upon their simple pay. Thus in 1855, the scurvy attacked the troops at Saint Omer in so serious a way, as to call for the presence of M. Maillot, a medical inspector; nor did it yield, until certain objectionable features in the provisioning were reformed. We uniformly observed, in Algeria as in France, that the troops employed in manual labor in the open air, in grading or stoning the roads, are in better condition, aside from the marked and conceded influence of physical labor upon the health; and this fact is explained by the extra pay which the soldier receives for these labors, and which in part goes to increase his food.

It is commonly said, that the soldier is better fed in the army than in his family, but this is not so true as is generally believed. Besides, ought not the amount of food to correspond with the sum of physical forces expended? The English laborers, who began our railroads, and whose unwearying vigor astonished our workmen, ate two pounds of meat daily; flesh nourished flesh. The French soldier receives, bone deducted, at most, one hundred and twenty grammes (a quarter of a pound) of meat in a day. I admit that he would eat less at home, especially if he came from a poor district, but at least he would have an abundance of bread, cabbage, beans, lard, and onions. Butter would vary his food; and milk, cider, or at least weak wine, are worth more than the water which he would drink in the army. His labor is free, and without disciplinary restraint, and he has no forced marches to make, carrying an equipment, which in a campaign weighs at least fifty-five pounds. He rests when he is weary, eats when he is hungry, and at night, instead of standing sentry, he sleeps a sleep that is sound, and breathes full volumes of fresh air, and not such air as is rationed out in the rooms of a barrack. The greater number of diseases, and especially of tubercular phthisis and typhoid fever, so common in the army, have no other cause than a vitiated air. In Algeria, the regiments which labored on the roads, during the heats of summer, in the open air, never had these diseases; but on the return of winter and bad weather, when they were driven into the barracks, they presently paid a heavy tribute to disease. The expense of supply for a company of eighty men is one hundred and sixty dollars per month, which is derived as follows: 'one hundred and fifty-six dollars and twenty-four cents, total of six cents and five mills per day, taken from the wages of each man; four dollars at least per month, from the sale of grease and slops, besides what is earned by the soldiers who labor in town, and that which is paid by the officers for orderlies, and which go to the soldiers in common. The company consumes daily in bread for soup, one dollar and twenty-one cents, in meat three dollars and thirty-five cents, in vegetables twenty-eight cents, in pepper and salt nine cents. This four dollars and ninety-three cents per diem, amounts to about one hundred and forty-seven dollars and ninety cents in a month; and allowing nine dollars and ninety-eight cents for washing, lights, and blacking, for brooms, and the barber's fee, there is left two dollars and twelve cents for some contingent expenses. Many commanding captains have a foolish tendency towards economising in the mess expenses, which results finally in large mortality. I have been astonished to see them intrust to a corporal the duty of buying provisions; as corporals are rarely insensible to the seductions of a glass of brandy; the merchants, who know their weakness on this point, turn it to their advantage, to the detriment of the company. It would be better that a special commission should have charge of the sustenance of the regiment. They could then deal directly with the producers, and by dispensing with intermediaries, benefit the regiment to the whole extent of the profits made by second and third hands. Bought upon hoof, the meat would be cheaper, and perhaps of better quality, the soldiers might cut it up themselves, and thus initiate themselves into camp life. It may be objected, that the soldier is suspicious, and ought to be left to dispose of the funds destined for his nourishment according to his own fancy; but the commission might embrace a representative from each company. Besides, the State, in supplying the funds, reserves to the fullest extent the right of control, and its interests are identical with those of the soldier.

CHAPTER III.

CAMPS AND SHELTERS.

THE three camps of the French army were located upon elevated grounds, with good hygienic conditions. The air circulated freely, and purified them constantly. The inclosure was, however, too scanty, the tents almost touching one another. There should have been left sufficient space to allow of their being moved occasionally, to purify the ground as it became infected by use. As for the barracks, the evil was without remedy. It is a fatal custom thus to huddle together the tents and barracks. In the Crimea, safety made this necessary; but at Constantinople, far from the seat of war, the camps, barracks, and the hospitals were altogether too close; and to the noxious exhalations thus engendered were to be attributed the persistence of cholera and the ravages of hospital gangrene and typhus. To the physician who asks for more space, it is replied that, it is of the first importance to economise the labor of the service, and so, to save a few steps, they violated some of the most simple, yet most important laws for the preservation of health.

The physicians also found that the location of the camps, even when nothing controlled the liberty of choice, was not always well selected. At Constantinople, a barrack camp was laid out scarcely half a mile from a miry plain, but the invasion of intermittent fever caused it to be abandoned. It is further to be observed, that two camps, or two barracks, or two hospitals, were never made upon the same model; and often a real perfection in arrangement was superseded by an injudicious innovation. It would, however, be sensible enough to prescribe a plan drawn up by a commission composed of officers of engineers and of the medical staff.

Camps, when stationary, tend rapidly to infection, but they cannot always be changed; for in winter the softened soil often prevents the work of removal; and, besides, camps often occupy military positions which may not be abandoned. We must submit to these necessities, with the determination of withdrawing at the first opportunity. The signing of the peace allowed us to remove our camps along the valley of the Tchernaïa, upon new and elevated ground, exposed to the sea breeze. The officers were loth to quit the conveniences they had arranged; but Marshal Pélissier commanded, and they obeyed.

When a camp ground cannot be changed, it is necessary to double our vigilance, in order to expel the organic miasms and purify the air, by sprinkling the soil in the tents with lime water, by placing a dish of hypochlorate of soda in a corner, by taking down the tent when the weather allows, or at least tying up the sides 80 centimetres (32 inches) from the ground during the most of the day. Soldiers take so little care of their health that they must be made to come out of their tents, or they will remain squat within, even in fine weather; and should be compelled to dry their clothes and blankets in the sun. The cavalry were more attentive than the foot soldiers (Zouaves excepted), to these simple requirements, which were issued as army orders. But the infantry gave them the most faithful trial.

The cemeteries were placed far enough from the camps to be free from noxious exhalations, and all the recommendations of the Army Council of Health were always regularly observed concerning them. Large quantities of quicklime and of chloride of lime, of which we had an abundance, were used in them and in the slaughter yards. It was said, and published, that the dead carcasses of animals poisoned the air of our camps; but this was untrue, as they were immediately buried.

General Canrobert, at the beginning, encouraged their burial, and this was regularly done.

The habits of cleanliness which distinguished the English army, should have been followed in our camps. They washed their body linen in warm water, and changed twice a week, but our soldiers were not so careful. Filthiness checks the functions of the skin, and engenders vermin. When a patient arrived at Constantinople, we first washed his garments in boiling water. On a review day, our soldiers presented, by their new and well-brushed uniforms, an irreproachable military aspect; but these fine battalions left, as they passed, the marked and well-known stench of the barracks. Is neatness incompatible with the soldier's profession? The Turk finds time, in the midst of a campaign, to perform many times in a day the ablutions required by his religion; and surely military discipline ought to be as imperative as the law of Mahomet! If that has achieved so meritorious a victory, military education should introduce gradually into the families of our laborers and peasants, those cleanly habits which we so envy in the English. This would be a national reform, which would result in profit to the public health. Our quarters for troops shine with the greasy filth of daily neglect. It is forbidden—would any one believe it?—to scrub the floors, the benches, and the tables, for fear of wearing them out. Why cannot a barrack be kept as neatly as a ship, and why cannot floors, waxed and rubbed by the soldiers, take the place of the imperfect tiling in the rooms? We have introduced, at length, this luxury into our military hospitals, in spite of the opposition of routine; and, upon entering the barracks, may ask with surprise, why such useful reforms are so slow in finding their place here?

The shelters of the army of the East were of different kinds, and for want of houses it adopted more primitive habitations. In the Crimea, we had huts, shelter-tents,*

---

* The *tente-abris* is a slight shelter, sufficient to cover a soldier in the bivouac, and so light that it can be packed up like a knapsack, and carried on the march. Its construction and use are described on a subsequent page. They are about being introduced into the American army.—TR.

and conical tents. The huts, which our soldiers called *mole-hills*, were dug a yard, at least, below the surface, and were about seven yards long by three wide and two and a half high. The floor and sides were covered with stone when they could be had. The walls were raised above the surface by brush woven together, and covered with a thick coat of clayey earth, and upon these was placed a roof, with double slope, made of the same materials. One or two openings in the roof admitted light, but these were closed with a sod when it rained. Whenever fuel was wanting, these huts were always dangerous to health, and the Piedmontese regiment, who lived in huts without chimneys or fire, had many sick.* On the other hand, the division cantoned in the forest of Baïdar could have had no better habitation,

---

* The pernicious habit of sinking the floors of tents below the surface of the ground has almost always been punished by increased sickness and mortality. Upon many occasions, while inspecting the camps in winter quarters, in the army of the Potomac, in the winter of 1861–2, the writer has observed that the sickly companies and squads were those that lived in tents with excavated floors. This disregard of health should be charged to the officers permitting it. The sides and floor of such a tent can never be dry, or the ventilation perfect, unless, perhaps, in cases where an open wood fire with a good draught is maintained. Carbonic acid gas will settle into the bottom of the tents, where the men lie, and typhoid fevers, rheumatism, and catarrh will swell the numbers attending the sick call and filling the hospitals. The practice prevents the frequent removal of tents to new grounds, and renders cleanliness impossible. The floor of a tent, somewhat raised above the natural surface, well drained, and previously dried by the burning of brush, presents the opposite conditions for the maintenance of health.

The army of the Potomac, in the winter quarters of 1861–2, were well supplied with wood, and very few tents were without fires. The fixtures for warming varied infinitely, and embraced a great variety of open fire-places, of brick, stone, mud, or sods, with chimneys of the same, with or without an additional draught made of boards, barrels, sticks and mud, or hollow logs; sheet-iron stoves, with or without pipes; open fires in the middle of large tents, with ventilation at the top; or, what was still more objectionable, kettles of embers from the kitchen fires. Of these, the open fire-place, with a chimney, was the only kind without its faults; and the chief difficulty in its use, arose from the increased quantity of wood required. The thousands of acres of oak timber, cut down as a military necessity, allowed no limit to be placed upon this luxury, except that of transportation.—Tr.

for being well supplied with wood, they kept up fires day and night. Nothing is so cheering as a large fire in a bivouac, and wood is half the life of a campaign. While warming himself in the open air, the soldier escapes the emanations of the common quarters; and when the fire is carried into the huts, it renews the air with great advantage. A day of reckoning with rheumatism may hereafter come; but in war we cannot always be so prudent.

I visited a Russian camp, where the troops live in huts similar to ours, but larger, wider, and deeper, with sheets of oiled paper instead of glass. Wood had become scarce, and the air, not cleansed by the fire, was heavy, damp, nauseating, and scurvy and typhus abounded.

To Marshal Bugeaud, we owe the ingenious contrivance of the Shelter Tent, made from the camp sack of the soldier. Its seams are replaced by rows of buttons, so that it can be changed at will into a square piece of cloth; and when two of these are buttoned together, and drawn over a pole, raised about a yard from the ground, and the corners fastened by pegs, the two owners of the sacks are sheltered under the tent. Thus was resolved the great question, as to how we might escape the double inconvenience of overloading the shoulders of our troops, and of transporting tents in the train of an army, by expensive and often impracticable means. Our troops become movable and nomadic, like the Arabs whom they pursue. This tent did us great service in the Crimea, but was of little service during the rigors of winter; for when placed on the ground it was too cold, and when covered with a bed of snow it was too warm, and the air within was rapidly tainted.

The conical tent is made for sixteen men. A single pole in the centre supports the top, while around, it is very firmly fastened to the ground by two sets of ropes, one of which is fixed, and the other movable, allowing the walls to be raised some two feet to ventilate the interior.* They resisted bravely the force of the winds,

* The Crimean tent has been furnished to a few of the regiments in the American army, but not so extensively as the Sibley tent, which is taller, and descends to the ground at a wider angle, covers less ground, and has a large opening for ventilation. The latter, resem-

and the Turks prefer them to all others. They make them of a very close woven texture, and the Sultan furnished a large number of very superior quality. Ours were made of cloth of open texture, which allowed the rain to sift through; but although not so warm, they were more healthy in summer, because they allowed the air to circulate, while in winter their fault was remedied by placing one over another.

Marquees are a more complicated affair than the conical tents, and as they do not well resist the wind they were not used for sheltering troops in the Crimea. Still they are more healthy and agreeable abodes; their vertical walls inclose a much larger volume of air, they are very easily ventilated, and are as roomy as a common chamber. On these accounts, they were used for the sick.* The English constructed during the summer, for their regimental infirmaries, marquees of large size, each holding twenty-four iron bedsteads, and as many bedside tables. The planks were movable, and kept with great neatness; and every patient had a mat by his bedside, and a hospital robe. An army on the march could not unfortunately enjoy a convenience so cumbersome; as to move one of these tents, with its furniture, would require at least twenty-five mules. In winter they were replaced by barracks. The Russians also employed tents of very large size for their sick, and I saw several erected near the hospital at Balbec; they served, so

bling in outline a Sioux lodge, from which it was perhaps modelled, was first supplied to our troops in the expedition to Utah, and possesses advantages over many if not over all other kinds now used. It is often stockaded to a height of three or four feet from the ground, thus affording additional accommodation. Around the central pole is placed the rack for arms, and a conical sheet iron stove with pipe passing through the ventilator, enables the inmates to keep their quarters comfortably warm in winter. In a cavalry camp, they may afford shelter to the saddles, an economical measure impossible in the common wedge tents.—Tr.

* The Marquee, or wall tent, covered with a fly or stretch of canvas, supported at the ridge by the same pole, but descending at a less angle, so as to allow of the free circulation of air between the two, is the variety furnished to officers in the American army. When two of these are united, they present an outer and an inner appartment, and very satisfactory accommodations.—Tr.

I was told, to isolate the soldiers tainted with severe and infectious diseases.

The choice of ground for a tent is of great importance. It should be exposed to the air, and free from dampness, should be upon elevated but not exposed places, and should have ditches for carrying off the rains. If, in winter, to keep out the cold, the tent is surrounded by a wall of dry stone, it should be taken down as soon as fine weather sets in. It is a great fault to bury tents to a certain depth to make them warm, for they are then damp, and more difficult to cleanse. In the Crimea, some of the tents had their floors muddy through the whole winter.

For a bed, every soldier should, under the regulations, receive a bundle of straw every fifteen days; but in a campaign this can seldom be furnished. It might be better to provide every man with a piece of waterproof cloth, which he could use as a shawl when it rained, and as a screen from the dampness of the ground in the bivouac at night. The sheepskin, which was given instead of the bundle of straw, became loaded with humidity, and propagated vermin. The field hospitals and regimental infirmaries had movable plank floors, and a kind of mat beds. After the taking of Sebastopol, some of the colonels covered the ground of the tents with wood brought from the city, or with a thin layer of nut tree branches, of which quantities were found in the forest.

The camp of the 81st regiment was a genuine model of arrangement. Its spacious tents were carefully in line, upon broad paved streets bordered with fir trees, which the regiment had planted. They were opened during the day, and contained a folding bed, which was turned up when not in use, and replaced when the hour for retiring came. The utmost neatness was preserved. Nothing was wanting. Even scrapers for the feet, made from broken sabres, were provided at the thresholds. In their infirmary, the regiment had, from its own resources, provided fifty beds; well arranged ventilators renovated the air, and an excellent fireplace kept the temperature at from 14° to 16° Centigrade

(58° to 60° Fahrenheit). Upon a visit, unannounced, I found the colonel, M. de Clonard, presiding at a distribution of oranges purchased for the scurvy patients; under a shed, I counted thirty or forty barrels of wine, kept in reserve for days of great fatigue. Fields of barley, wheat, and potatoes had been sowed for the common supply, and they had even constructed, at the camp, a plough of the Dombasle fashion! The regimental band daily discoursed cheerful airs upon a beautiful esplanade, planted with trees by the soldiers, and adorned with a fine rustic café. Beyond the color-line there was a row of little stone buildings; the boxes that had contained the preserved vegetables furnished materials for the roofs, and even for stovepipes. These were the kitchens of the several companies. M. de Clonard had thus turned to account the thousand pair of hands of his regiment, when war gave them no further employment; had banished homesickness and diseases, introduced gaiety and health, and preserved his effective strength almost entirely unimpaired.

The English army passed the whole of the winter of 1856 in well closed barracks. Every morning, the boards upon which the soldiers lay, were sprinkled, with fine sand, which was swept off at night; stoves of mineral coal were kept continually burning, which allowed of the ventilators being always open. Two temporary barracks served as reading-rooms, where there were books, benches, a table, pens, ink, and paper. The English soldiers burned the offal of their camps, while the French buried it. In winter, the heap of offal burned with difficulty, and with a black, stinking smoke that spreads through the whole cantonment.*

The materials for building, taken from the ruins of Sebastopol, having been divided equally between the English and the French, were then apportioned to the regiments. Without these, the army would have suffered cruelly during the winter of 1856. It was worth

---

* The English reports show that their manure heaps were swept up, and, on the whole, well burned. The most approved manner of burning was in kilns, made somewhat after the fashion of lime kilns.—TR.

noticing the zeal with which the soldiers sought wood under the rubbish, and loaded it upon their shoulders or upon their *arabas*. Planks, beams, windows, broken doors, bricks, tiles, and, in short, everything was taken that could be turned to use. The Russians, seeing them so busy, sought to annoy them with cannon, but our soldiers allowed no such little affair to disturb them. I have seen them climb upon the roofs of the highest buildings, to strip off the sheets of zinc. The Russians would shoot at them as at a target, and they would reply by a mocking gesture well known among the blackguards of Paris.

CHAPTER IV.

CLOTHING.

As, in the Algerine war, we introduced into our military costume, modifications appropriate to the climate, so, in the Crimean war, we borrowed from the native Tartars certain garments which better shielded our soldiers against the rigors of their winter.

The *Criméenne* is a long and ample hooded cloak, with a little cape, and falls to the middle of the leg. The cloth is coarse, but warm, and almost. water-proof. Excepting the general officers, who wore an overcoat trimmed with fur, everybody wore the *Criméenne*, and it replaced the African *burnous*, and the *Caban*. It proved very useful, and will perhaps be regularly adopted, as it guards the soldier from diseases acquired so often by passing suddenly from the high temperature of the guard-room to the cold outside air, in mounting guard at night. The hood shields the head and neck from the cold, the wind, and dampness; prevents the engorgement of the glands of the neck, and the bronchitis, to which they are liable from the chill. A preparation of India-rubber would easily render the little cape which covers the shoulders water-proof. This garment would replace with advantage the blanket, which the soldier carries upon his knapsack, and which gives him so ridiculous an appearance.. The blanket, so awkwardly perched upon the knapsack, when wet, is very heavy, and dries with difficulty. When dry, it weighs about three pounds and a half. Therefore in summer, in order not to load the shoulders of the men too much, they give them only a half blanket, the other half being kept for the approach of winter. The storage of these half blankets is not easy, and the army runs the risk of being deprived of them if their supplies cannot follow them. But the *criméenne* has none of these inconveniences, is not so heavy, and can be made still lighter, and the knapsack is relieved from the difference in weight.

French taste, which sometimes blunders, has often attempted to rob this garment of its essential qualities. To render it more elegant, the officers have worn it shorter, narrower, and without the cape or hood, but it was no longer the *criméenne*, and had lost its peculiar advantages. The only change which appears practicable, would be to place a flap behind, as in the old infantry cloak, with the view of giving them at will more or less breadth, without drawing in those great flowing folds, which draped so nobly our brave soldiers, and gave them so majestic an appearance. The *criméenne*, with the tunic, would form the winter dress. The tunic appears to me scanty, and should be made more pretentious. A loose uniform is at the same time more healthy and more military; the costume of the Zouave is an example.

The Russian officers and soldiers wore a grey habit, very coarse, but warm, and shedding the rain very well. It descended almost to the feet; strings to allow of its being drawn in at will gather the folds upon the back, and this gathering does not give them a very graceful appearance. I would much prefer the flap which we had, and which the Austrians still use. The habit of the officers, and even of the generals, is like that of the privates, except a little lace upon the shoulder. The lace of the generals is ornamented with two or three stars, according to the rank.

The English have not adopted the *criméenne*, but have borrowed from us the tunic, and have adopted as an overall, a long spencer, of brown knit stuff, protecting effectually the chest and lumbar region, while it leaves a perfect freedom of movement; but this does not compensate for the qualities of the *criméenne*. Our allies have supplied its place with several distinct garments. In place of the hood, they adopted a kind of otterskin helmet, falling over the ears and cheeks, and leaving none of the face in sight but the eyes and mouth; and the cloak was replaced by a large piece of india-rubber cloth, upon which they lay at night in the bivouac. Our troops also wore, at the beginning of the war, a kind of spencer, with sleeves made of sheepskin, the wool within, and next to the body. This costume was not graceful, but what was worse, it often was too warm, and caused profuse perspiration. When the cold weather was over, there was danger in leaving it off, as the body had been accustomed to the moisture. The wool retained the humidity, and the grease defiled the outer garment, while vermin found ample shelter. It was entirely given up.

A girdle of flannel is the best preventive against diarrhœa, the precursor of dysentery, so fatal in armies; and the old soldiers, accustomed to it in the African war, took care not to leave it off. The conscripts, who knew not its value, lost it, or left it in their knapsacks, in which cases the blame ought to rest upon the officers and physicians; as the measure was prescribed by the minister of war, and they ought to see it carried into effect.

Each English soldier had two flannel shirts. Nothing is more healthy than woollen, for in winter it gives a genial warmth, and promotes the functions of the skin, while in summer it prevents the arrest of perspiration. The Arabs scarcely wear any but woollen garments, and our marines use them in all climates. Two woollen shirts are scarcely heavier than a common soldier's shirt, and could take its place in the knapsack. When wet through, the soldier by putting the other on might escape bronchitis, which is so common, or pneumonia, which proves so often fatal. While I await the adoption of the woollen shirt for our soldiers in the field, I ask that it should be given to our sick, in the regular and field hospitals.

The whole army was provided with long gaiters made of coarse warm cloth, and reaching to the knees, like those of the soldiers under the Empire. They supported the leg sufficiently in marching, and prevented varices. The greaves of leather actually used, hardened by wet and by frost, excoriated the ancles, and were too cold in winter, and too warm in summer. On the other hand, those of cloth, besides wearing out sooner, retained the wet when it rained, and in fact acted upon the leg like a sponge. The soldiers, having but a single pair, could not always be dry. They also had greaves of sheepskin, the wool on the inside, but they also retained the moisture, and, dried before a fire, became horny, rigid, and brittle. It was not an easy thing to recognise the soldier, in a man wearing a spencer, sheepskin greaves, and wooden shoes. But this outfit was only a temporary one, and was soon put aside.

Woollen stockings, although excellent when dry and clean, often remained wet when the soldier was deprived of fire in the bivouac; and many a man has had his feet frozen, by sleeping with wet woollen stockings and shoes. The woollen slippers contained in the wooden shoes, were always dry; and the latter were often necessary, even in very cold weather, when the shoes, long damp, had been so hardened by the cold, that they could not be put on until they had been softened by warmth. The common shoe, covered with a greave of leather, was not sufficient in a country without roads and thoroughly sodden with wet. After the sad experience of the winter of 1855, the English furnished their troops with strong boots of yellow leather, soft, waterproof, and reaching to the knees, such as those used by sportsmen. They were an aggravating luxury, and half boots were even preferable upon the march.

The Russians, who understood the country, had adopted the half-boot. The firmness and closeness of their leather enabled them to travel in underbrush woods without tearing, or in the water without soaking them. The leg of these half-boots was large enough to admit the pantaloon legs. This kind of shoes ought to be adopted for our soldiers, two pairs being furnished to each man; one for summer, with greaves of white cloth, and the other, a half-boot, for winter. Remembering with what eagerness our soldiers despoiled the Russian dead of their half-boots, we may be assured that this reform would be popular with them; they know what is good for themselves.

In times of peace, the knapsack contains articles for the toilet, and four packages of cartridges without balls; the total weight is fifteen and a half pounds. If we add the blanket, shelter tent, little wooden bowl, little tin can, cross-belt, waistcoat, cartridge box, sabre, gun, and bayonet, and two days' rations, we have about fifty-three and a half pounds weight. In a campaign, this is considerably increased, and sometimes amounts to sixty-six pounds, at the time of departure. In the African wars, the soldiers carried six packages of cartridges, and provisions for eight days. Furthermore, the large canteens, large bowls and kettles, must be carried by a detachment of eight men.

It will be admitted, that opportunities were afforded for studying these details in the Crimea, when it is remembered that cold weather was just coming on, that the army had suffered greatly during the previous winter, and that endeavors should be made to spare it as much as possible from new exposures, by turning to account past experience and the counsels of hygienic science.

Having made this research into the best means for preserving the health of our army, I reduced my views into the following report, addressed November 10th, 1855, to the Minister of War:

"To His Excellency the Marshal.—My duties have been actively applied to the threefold question of food, shelter, and clothing.

"It is true, our army has no longer to endure the miseries of the trenches; but instead of old soldiers, it consists at this time of one-third if not a-half recruits; young beardless soldiers, with at most but a year of service. Last winter, they each received daily an extra pay of half a franc on account of siege work. As this resource will be cut off, the daily allowance will suffer to that extent. The army was fed last winter with great care, but still the number of sick, for the first five months of the cold season, was high.* It is highly important to remember, that the army is now much more numerous than then; it numbers at this time over 140,000 men.

"1st. *Of the Shelters.*—To the troops encamped in the forest of Baïdar, I advise the building of huts buried a yard and a half in the ground, with roofs of double slope made of brush, covered with earth, or better still, sodded. At the bottom of the room there should be a fireplace opposite to the door, which should be constantly fed with wood from the forest, to renew the air, especially in its lower strata, to dry the walls, and to render a habitation that otherwise would generate typhoid fevers and scurvy, a warm and healthy abode. Where wood and water abound, the soldier is happy. Instead of carrying bread to the distant cantonments of Baïdar, we sent sacks of flour, which was made into bread on the ground, thus economizing

---

* The following statement shows the sanitary condition of the French army in the field hospitals of the [Crimea, in the winter of 1854–5:

Months.	Effective Strength.	Sick.
October	46,000	3,200
November	55,000	5,000
December	65,000	6,000
January	75,000	9,000
February	88,000	8,000

These figures do not include the sick in the regimental infirmaries. The hospitals at Constantinople presented during the same period the following conditions on the 25th of each month:

Months.	Sick present.	Sick.
October	3,235	1,447
November	3,486	2,695
December	4,414	2,427
January	7,031	4,084
February	7,386	4,905

AUTHOR'S NOTE.

the wood which came to Kamiesch from Varna. This forest, in a hygienic point of view, fulfilled expectations. The six weeks which three divisions of the first corps spent there, could not have been more favorable to the health of the troops, and especially that of the recruits.

"The camps placed upon the undulating plateaux of the Crimea, were also in perfect health. Unfortunately not a single tree was left; and the subterranean forest, that is, the roots of trees cut down the year before, are nearly exhausted. It is useless to think of building huts, tents must be resorted to. Where the soil was calcareous, they dug a circular pit, some two feet deep, in which the tent was placed, making a gutter around it, for drawing off the rain water. The materials taken out served to build a wall around it, about two feet high, so that the soldier, when in bed, was sheltered entirely from the wind and the rain. The shelter would have been complete had they added a fireplace, as in the officers' tents. Where the soil was not calcareous, the arrangement was sooner made, but not as good, for the circular bank, in form of a parapet made around the tent, was not to be compared with the wall of dry stones, and the ditches had to be paved, to hinder the water from filtering into the inside of the tent. It is necessary to furnish the men with either a sheepskin or a plank (biscuit boxes answer the purpose), to keep them from the ground, and a piece of oiled cloth which they could form into a mantle, by wrapping it around them on rainy days.

"The shelter-tent is entirely insufficient for winter, and it is so short that it does not cover the feet of the men. It may be advantageously replaced by the conical tent, fashioned after those of the Turks, of all tents the warmest and strongest to resist a gale of wind. Tents should always be set as far apart as possible, and when the weather permits, should be moved at least every four days. When the sun shines the contents should be exposed to the air, and the tents should themselves be taken down; but unfortunately this very essential requirement is not attended to, even in the field hospitals. It should be signalled by tap of the drum, which never fails to catch the ear of the soldier.

"Board barracks would be better than tents, when the joints of the roof boards are tight, but this shelter is exceptional and limited entirely to the sick in the regular and field hospitals, and the regimental infirmaries.

"If we help the soldier a little, he will help himself. Stones abound everywhere, and of them the four walls of a house may be easily built, and we have only to furnish boards for the roof, to have quickly constructed, and at small expense, a range of comfortable houses, to shelter the men from drenching rains, and to dry their clothes when wet. Without a fire, they will wear their wet clothes more than a week, and this causes many diseases, which must be expelled from the army. When stone cannot be had, I would advise that the walls be made of gabions, sand bags from the trenches, butts, and biscuit boxes.

"For three days, there have been distributing to the army the building materials found in Sebastopol. Thus use is made of the town, which is being gradually wasted by Russian cannon. I examined the materials with care, and found a large quantity of plank, carpenter's wood, and tiles, besides a great number of large iron kettles, which we shall use for making soups for the hospitals. These resources, apportioned discreetly, will be of great benefit; and within a month, if the fine weather should continue, our camps, which are the scene of wonderful activity, will be completely transformed.

"2d. *Of the Clothing.*—The Crimean cloak has been of great service, and it is urged that it should be furnished to all the soldiers. They have nevertheless done wrong by wearing it in summer, instead of keeping it for bad weather in winter. This abuse of it renders them sensitive to the cold, and exposes them to the effects of vicissitudes. The flannel girdle is indispensable, in preventing and checking the diarrhœas that are so common, and that so often run into dysenteries and other very serious maladies. It should be applied in direct contact with the body. Our old soldiers know its usefulness, but it is not easy to make the recruits wear it. I invoke for the enforcement of this regulation, the vigilance of all commanding officers and regimental physicians.

"In winter, we distribute to the troops another half-blanket, to be used with that given them for summer use. This blanket loads down the soldier on the march. When it becomes damp, as it does by the first rain, it can hardly be said to get thoroughly dry the whole winter. I am convinced it could be profitably replaced by a red woollen shirt, like that worn by the English. A woollen shirt keeps up a pleasant and uniform warmth. Every man should have two, which, at 75 cents each, would amount to $1.50, or about the price of a blanket. The men would be less loaded, and they would have next to the skin, a warm, dry, and perfectly healthy garment. Flannel shirts ought to come into general use in our infirmaries and field hospitals, where they would prevent and cure many diseases.

"The wooden shoes, which our soldiers use as a change for their wet shoes, are indispensable in a country where the ground is trampled up to a considerable depth; and during the last winter, the men whose shoes were frozen hard for many days, could not have gone out had it not been for these wooden shoes. Socks are very useful, and not only supply an indispensable outfit for walking in wood, but are also of precious use during the night, in keeping the feet from freezing. It might be difficult to supply them for the whole army, but General Bazaine has assured me, that in every company there may be found men who will knit them for their comrades for a very moderate price. Besides, with some of those ingenious machines of which models were shown at the great Paris exhibition of 1855, we might be readily supplied.

"3d. *Of the Food.*—We cannot too highly praise the department of subsistence for having so happily solved the difficult problem, of provisioning an army eight hundred leagues from France; and at no other period in our military history, has the daily issue of rations been made with more regularity. It did not fail a single day, and the alternation of fresh bread and biscuit, of coffee, wine, and brandy, and of fresh meat, preserved meat, and lard, was conducted with facility, breaking the uniformity of food, and resulting in general health.

"The supplies most needed were fresh vegetables, and to this want, as well as to the cold and damp habitations, and to sleepless nights spent in the trenches, was to be ascribed the scurvy, which so seriously embarrassed the army. To supply the want of fresh vegetables, an abundance of preserved mixed vegetables should be provided, sourkrout, potatoes, and onions; they are the best for the soldier's use. Seeds for sowing culinary gardens, and especially radish seeds, should be distributed to the companies, and it would be desirable to supply the mess with condiments, such as cloves, long pepper, nutmegs, and laurel leaves. Thyme abounds here, and I advise its use in seasoning soups. Cargoes of oranges and lemons, sent to the Crimea, will be necessary in treating, and even in preventing, scorbutic affections. Of vegetable acids the army of the East had long been deprived."

I addressed a copy of this report to Marshal Pélissier, and to M. Blanchot, the intendant-general of the army. In a reply received from the latter, he said:—"I observe with pleasure, that most of the hygienic measures advised by you have been executed, and we have gone even farther than you require, in regard to clothing. You seem to regard it as difficult to furnish the whole army with socks, but I am happy to inform you, that in the coming winter every soldier will have not only a pair of socks, but likewise a pair of woollen stockings, and a pair of long gaiters." It will be seen that my hygienic views agreed entirely with the plans of the intendant-general of the army. The result of these studies will show that my medical and surgical suggestions were also uniformly sanctioned by the Minister of War, and by the Marshal commanding in the East. We can never form too high an estimate of the services which medical science can render to an army in the field, and of the influence that it may exert upon the vicissitudes of war. Its counsels, which are not always asked or heard until suffering and death make us cruelly feel their value, might have saved many a man who has lost or imperilled by imprudence, a life of which the nation had need. The preservation of the soldier, sent out at great expense, is the first thing of interest to a people who may be conducting a foreign war, and it is the first pledge of success. Diseases slay more men than steel and powder, and it is often easy to prevent them by a few simple hygienic precautions.*

* The history of the United States has hitherto afforded but slender materials for statistical deductions of the sickness of armies, and of losses by war. In the Mexican war, the standing army was increased by ten new regiments, and a large volunteer force, which was mustered out of the service upon the peace. The following summaries show that there occurred most extraordinary percentages of losses from disease, as compared with those by death or disability from battle:—

	Old Establishment.	Ten New Regim'ts.	Volunteers.	Total.
Aggregate officers and men..........	15,736	11,186	73,532	100,454
Average length of service (months).	26	15	10	
Discharged by expiration of service.	1,561	12	50,573	52,146
" for disability.............	1,782	767	7,200	9,749
" by order of civil authority	373	114	2,016	2,503
" total.....................	3,716	893	9,216	13,825
Deaths, killed in battle, officers....	41	5	47	93
" " " men.......	422	62	467	951
" died of wounds, officers.....	22	5		27
" " " men.......	307	71	100	473
" total killed and died of wounds, officers..........	63	10	47	120
" " " men..........	729	133	567	1,429
" ordinary, officers..........	49	86	} 6,256	10,885
" " men..............	2,574	2,055		
" accidental..................	139	30	192	361
" aggregate..................	3,554	2,264	7,078	12,896
Wounded in battle, officers..........	118	36	129	283
" " men............	1,685	236	1,189	3,110
Resignations.......................	37	92	327	456
Desertions.........................	2,247	602	3,876	6,725

TR.

PART II.

THE FIELD HOSPITAL AND MEDICAL SERVICE.

Our army of the East had three kinds of health establishments, corresponding with the three grades of treatment. The infirmaries and ambulances of the trenches, were the first to receive the sick and wounded; those who were suffering more seriously, went to the field hospitals of Divisions; and lastly, in addition to these, were the regular hospitals, placed beyond the seat of war, and receiving the sick that required a long and careful treatment. Having begun my inspection in the Crimea, I first had to notice the infirmaries and field hospitals, and my first observations were directed to the Surgical Service.

CHAPTER I.

THE INFIRMARIES AND FIELD HOSPITALS.

WE were unable to establish definitively the regimental infirmaries, until after the taking of Sebastopol. Until then, the temporary nature of the bivouacs prevented their proper arrangement, and only the parks of artillery, and the engineers, were settled and had their infirmaries under barracks. That of the artillery park, at head-quarters, embraced everything that could be desired, and was ably conducted. It had around it a vegetable garden, exclusively reserved for the sick, thus improving and varying their regular allowances. This infirmary sent but few sick to the field hospitals, or to the regular hospitals.

The recruits who arrived during the autumn of 1855 had to undergo at the same time a change of climate, a new kind of life, and a rigorous winter, and it was to be apprehended that they would furnish many inmates to our hospitals. In view of this unhappy prospect, Marshal Pélissier prescribed and allowed to each regiment, two barracks for infirmaries. ' I would have wished that, to prevent crowding, the important measure had also been adopted of sending to Constantinople fifteen thousand of the soldiers, who were ailing, and too accessible by disease, they would have then passed a mild winter in comfortable quarters ; but it was impossible to act upon my suggestion. At least the reorganization under barracks of infirmaries forty beds to a corps, served to furnish the army with new shelters for twenty-four hundred sick.

The internal arrangement varied according to the regiments, and while some of the barracks were chinked up and very close, others were open to the day, the joints between the boards admitted the rain, and although the stoves were always burning the cold was intense. The corps of engineers was blamed, as if it could do everything. It had built the barracks, and closed the joints with battens, and if the dryness had drawn the wood apart, the inmates should have repaired it. In some infirmaries, the sick had beds made of strong cloth stretched upon wooden frames, or bundles of boughs covered with little mats, but all the others were reduced to the dirty plank of the camp bed. Most of the barracks were whitewashed with lime within, and disinfected with chlorine, but these healthful measures were sometimes neglected. The diet presented the same irregularities, and while in some a small amount of funds reserved from the wages for labor allowed of improving and varying the food, in others, nothing was changed from the ordinary regimen of the soldier, the modification being only a diminution of quantity. Cleanliness was everywhere wanting. Such an indifference is truly incomprehensible. In each infirmary there were fifteen or twenty men, limping, listless, and idle, and yet they were not employed to clean their own room! And negligence was tolerated which endangered the health of the sick! Would it not be practicable to inspire more system in the hospital service throughout, by leaving a large part to the watchfulness of the colonels, who might be aided by rules carefully prepared? In a campaign, without doubt, we must do as we can, for our means are often scanty, but still we should see that what is indispensable should never be wholly wanting.

In two or three infirmaries only, I found a special register giving the names of the men of the corps who had been wounded by the enemy since the beginning of the war, and the date, the place, the severity, and the result of the lesion. It is to be regretted that this example was not more generally followed, and that it was not prescribed as a regular requirement. The authenticity of these documents would render them very useful for statistical and other reports. It would be the golden book of the regiment, its titles of nobility.

The good arrangement of infirmaries is of great importance. Being the first asylum of the sick and wounded, they should send to the field hospitals of divisions, or to the regular hospitals, those who require a long treatment. In those badly arranged the most simple ailment may become severe and degenerate into a long course of disease. In a healthy climate like the Crimea most of the diseases are at first light, and may be checked at the outset by a little repose and careful attention; but if these hygienic precautions are inadequate, the field hospitals become encumbered. To check an ailment in its early stages by applying immediately the first remedies; such is the use of the infirmaries. As for the wounded in battle, the first dressings are almost always applied in the ambulances of the trenches.

Some eighteen hundred yards from Sebastopol, we find hidden in a snug little valley, a little farm-house of humble appearance, which I never could look upon without respectful emotion. This place had served as the ambulance of the trenches for the works of the left division. Established when the siege began, at the far renowned house called the Clocheton, the field hospital was forced by the enemy's fire, which was unceasing, to retire into the nook where this little farm-house stood; the bearers of litters brought in the wounded continually, and on the night of May 1-2, 1855, four hundred entered. As the works of the siege advanced, the hospital spread itself; tents and barracks grouped round the original building. A worthy chaplain from the fleet resided with the physician. Religion and science united to solace the pains of the wounded, to inspire a hope of life, or soften the hour of death. A piece of ground inclosed with walls, was used as a burial-place, where each officer had a separate grave, while the soldiers reposed together in great common graves;—companions in arms and in dangers, death itself did not separate them. After the fall of Sebastopol, this field hospital became a shrine of pilgrimage, where each one sought the tomb of a friend. Well might a chapel be erected there in honor of so many courageous men, struck down without distinction in the painful labors of the siege.

Two caves used as ambuscades by the Russians, in the ravines of Carénage, and of Karabelnaïa, served as field hospitals for the trenches of the right. They were sheltered from balls fired in direct range, but more than one bomb rolled down the ravines, exploding and making victims at the doors of these sad retreats, the habitations of suffering. Only a straggling light penetrated these recesses by day, rendering surgical operations difficult; and at night, only one small lamp was suspended from the roofs so as not to attract the notice of the enemy. In the midst of the continued and deafening thunders of the cannonade, might be heard at intervals the cries of the birds of prey, as, disturbed in their wonted homes among the cliffs, they plunged down, and bore away the rags of human flesh which were scattered around. After the taking of the city, we went with pious reverence to visit these grottoes, peopled with so many gloomy memories. They showed us the litter of straw, still bloody, where the surgeon had knelt to extract a ball, or stop a hæmorrhage. Who can ever tell all of the sad and pitiful scenes which these places have witnessed!

4*

In these ambulances of the trenches, gaping wounds and broken limbs received their first dressings; the blood which flowed in abundance, was checked by hasty appliances; many entered only to die, after cruel agonies, with heroic courage, while the remainder were transferred to the field hospitals of Division.

The fourteen Divisions of the Crimean army were each provided with a field hospital, but they sometimes simplified the labor, by giving one only to two Divisions, and in this case it was always a double size. The Divisions were arranged into three corps, to each of which was attached a head physician. Each field hospital had eight doctors, two majors, and six aid-majors. The apothecary service was intrusted to one or two military apothecaries, under one faculty.* The number

* The medical officers of the French service rank as follows in the descending series, with the total number in the army of each grade.

    7 Medical Inspectors.
  40 Principal Physicians of the first class.
  40 Principal Physicians of the second class.
130 Major Physicians of the first class.
260 Major Physicians of the second class.
400 Aides Major Physicians of the first class.
400 Aides Major Physicians of the second class.
300 Sub-aide Physicians.

The apothecaries are divided into like classes, with corresponding names, but their numbers are less, viz.: 1, 5, 5, 17, 34, 50, 50, and 160 respectively. These numbers are alike in peace and war, and they bear no corresponding relation with military grades. Physicians and apothecaries from civil life are commissioned as auxiliaries for special emergencies according to the wants of the service.

The functions of the several grades of medical officers are defined as follows:

The Medical Inspectors, designated by the Minister of War, acting collectively, constitute the Council of Health of the army. The Inspector of oldest commission presides, and the Board is charged with overseeing and directing all that pertains to the healing art, in every branch of the health service, and with assisting the Minister of War upon all subjects to which these questions relate. They maintain a correspondence with the medical officers of hospitals and corps of troops, and with the chief health officers of the armies, in all that concerns the science and art of healing. They advise in the designation of medical officers in the various departments of the sanitary service, upon information derived from the notes or reports of the persons engaged in the two professions, and from the evidences

of infirmaries increased in proportion with the sick, and the medical service was apportioned according to their wants. When a part of the division made a movement, it was followed by a part of the field hospital, under the care of the second major physician, and two aids. The stores carried upon ammunition wagons or the backs of mules, were greater or less in quantity, according to the probable wants, and the facilities for transportation. They were often compelled to leave behind the ambulance caissons, and to march with only a few chairs or litters, to bring back the sick and wounded.

At the opening of the war, the field hospitals of Divisions were in tents. The barracks arrived at a later period, and became more and more numerous. At the

shown by the examinations of the major physician, undertaken with the view of ascertaining special aptitude for the functions of the physician or the surgeon. Individually, the Inspectors are charged with the duty of annual or extraordinary medical inspections, under special instructions. They may be employed in the direction of the medical service of the armies, and upon such special missions as the Minister of War may assign.

The Principal Physicians are attached to the corps of armies in campaign, and perform for their commanders, duties analogous to those of the Council of Health to the Minister. They are employed as chiefs of the medical service, in hospital establishments whether civil or military, and in this respect their functions are determined by the regulations of the hospital service.

The Major Physicians, of the first and second class, are employed as practising physicians in hospitals, and as chiefs of the health service in corps of troops. Their fitness for this service is previously ascertained by examinations, of which the programme is arranged by the Council of Health.

The Aides Major Physicians have their two classes, each divided into two sections, in order to facilitate the rotation of these officers in the hospitals and regiments, so as to prevent them from remaining too long in any one part of the service. They have the immediate charge of the sick, and their duties nearly correspond with those of surgeons in the American service.

The Sub-aide Physicians, like the Assistant Surgeons of the U. S., render medical assistance in the regiments and hospitals, and are often placed in charge of detached bodies of troops, and small posts.

The Auxiliary Physicians, commissioned by the Minister of War, or employed by the Military Intendants, are employed in the duties of Aide-Major, and cannot be intrusted with those of Major Physicians, except in default of such officers in the service.—M.

close of the year 1855, they could lodge from four thousand to five thousand sick. At that time, the Minister of War sent six thousand new mattresses, an immense quantity of blankets, beyond even their wants, and a considerable amount of stores. It was not easy to take care of all this in a campaign; every moment, unforeseen difficulties arrest the best intentions; thus during the most rigorous part of the winter of 1855–6, it was impossible to wash the bed clothing properly. To lessen the great consumption of linen bandages, it was desirable they should be washed, so as to be used again, but there was no laundry. It became necessary to burn them, so as to preserve the hospital from the putrid emanations which they gave out. The surgeons in the field should be economical in the use of lint and linen compresses, the supply becomes more difficult to renew in proportion as cotton is more generally employed as a substitute for linen and hemp. Carded cotton increased our resources, and a compress of wadding maintains a pleasant and uniform heat; it is an excellent dressing, and well to recommend. It is true, wadding is not an absorbent, but this inconvenience is obviated by placing between it and the wound some tufts of lint.

The diet in the field hospitals of Divisions was essentially the same as that of the hospitals in France, except such accidental modifications as were imposed by unforeseen events. Broth was never wanting, and besides meat and vegetables, the field hospitals received minced pies, eggs, prunes, sweetmeats, chocolate, and wine. They were even furnished with cans of preserved milk of the consistence of butter, which was diluted with three or four times its volume of water when used. It retained its properties, even after several days' exposure to the open air. The physicians sometimes distributed Bordeaux wine, derived from the national gifts. These field hospitals were of course moved many times during the war, and the following history of the field hospital of the 3d division of the 2d corps will serve as an example of the rest.

On the 20th of September, this field hospital was divided at Alma into two parts, of which one was established on the battle-field, and the other in the train of the Division. Four hundred wounded, of whom a hundred were Russians, were attended to during the night, and at once embarked. Fifteen amputations were performed. Arriving under the walls of Sebastopol, the field hospital was established in the ruins of a Tartar farm-house, and received the first wounded of the siege. On the 6th of November, the day of the battle of Inkermann, a section was detached to go to the camp of the Moulin. It received 400 wounded Russians, and as it was easier to bring tents than remove the wounded, they remained for the time on the field of battle, with a brigade of the Division. They had pitched temporarily upon the place of combat, without regard to choice of site, and bad weather subsequently prevented them from removing it elsewhere, or even from enlarging it. It was upon a flat piece of ground, sheltered and hedged in on every side by the French and English camps; but it was near the siege works, at the entrance of the ravine of Carenage, and near that of Karabelnaïa, and into it the ambulances of the trenches emptied the greater part of their wounded. After one of the night combats, which were so murderous and so frequent during the winter of 1855, one hundred and thirty great operations were performed within twenty-four hours.

Meanwhile, the 3d Division of the 2d corps had sustained many cruel losses; its most valiant chiefs had fallen at its head, and its effective strength was reduced to 3,000 men. It received orders to move upon the Tchernaïa, and there replace the 1st division. Each Division left its field hospitals and its sick in their respective positions, the medical staff and hospital staff being only changed, by passing from one field hospital to the other.

Thus, the field hospital of which we have spoken, changed, so to speak, its Division but not its locality. Scarcely a hundred yards separated it from the reserved artillery parks of the French and English. On the 16th of Nov., 1855, at about four o'clock in the afternoon, three fearful explosions, echoed from the mountains in a prolonged roll, shook suddenly the soil of the Crimea, and announced some terrible accident. A high column of flame and smoke rising in the direction of the mill of Inkermann, marked the spot where three magazines of our grand park of artillery, containing 66,120 pounds of powder, 600,000 cartridges, 300 loaded bombs, and other appliances of war had exploded. The burning materials, thrown to a great distance, set fire to the English park near us, and occasioned partial explosions and a great conflagration. We had 100 wounded and 30 killed, while the English lost 24 killed and 112 wounded. The misfortune would have been still more serious, had it not been for the vigilance of Marshal Pélissier, who directed in person all the measures for safety. The field hospital barracks were thrown over like a pack of cards upon the poor sick, but happily only a few were bruised. Five physicians in the field hospital were hurt, so as to be incapable of performing their duties, but nevertheless, during the evening, all the wounded had received the necessary care. The sick in this field hospital which was not rebuilt, were sent to another hospital. The field hospital which the 1st Division left to the 3d, was in a favorable situation upon the plain of Inkermann. The soil was dry, elevated, gently inclined, and exposed to the constant ventilation of a healthful breeze. It was an elongated quadrilateral, divided into two equal parts by a paved road, the entrance guarded by a sentinel, who occupied, somewhat after the manner of Diogenes, a hut made of two butts, set one over the other, and open on the side. The inclosure was formed by a trench, and a pile of butts filled with earth, served as a parapet. The shelters consisted of marquee tents, single or double, and of Turkish tents or else barracks. Of the twenty-four barracks seventeen were provided by the English at the beginning of the campaign, but they were not so good as those the army subsequently received. Low, damp, and badly ventilated, they were used only from necessity. Those applied to the use of the physicians were in the midst of the hospital, where after exposure during the day to miasmatic affections, they remained at night still exposed to danger. We cannot too strongly insist upon the peril and folly of such imprudence. These officers almost always exceed their duties, and remain in the hospitals after their service is finished, even refraining in times of epidemics from taking horse exercise, and thus neglecting those preventive measures which they recommended to others. This excess of self-denial deprives the army of competent men, and endangers the attendance of the sick. There could be nothing to prevent the officers of health from lodging at least 200 yards from the hospitals, leaving in them only the medical guards during the night.

It should be remarked that, at times, the medical staff was so much pressed that human endurance and the most zealous activity could not supply the requirements. However numerous, in time of battle or of an epidemic, they proved very inadequate to our wants. When a half day's battle sends to a hundred physicians in the field hospitals six or seven thousand wounded at a time, could they even place so much as a single compress and bandage upon each wound? Much less could they perform upon each the operations indicated by surgery.

To supply this want there was created, in the Crimea, a class of attendants of subordinate grade, who rendered very important services. Our system of recruiting makes our army the vivid image of our society, and assembles under the flag its manifold elements. Among the convalescents were often found men of education, bachelors of arts, and even lawyers. Some of these returned to their families on sick leave, but we retained those who appeared capable of assisting the physicians. These new duties, by employing their minds, hastened their recovery, and some, becoming fully restored, returned to their corps, to be replaced by others. These useful auxiliaries are called *soldier-dressers*. Scrive, Thomas, and Lustreman, in the Crimea and at Constantinople, were warm in their praises of their promptitude and skill. When the typhus decimated our medical corps, we feared, for a time, that we should be left without physicians, and urged.

the Minister of War to send some as soon as possible; but he had none at his disposal, and recruiting failed to supply the want. Thanks to our soldier-dressers, we triumphed over this serious difficulty, but without them, our medical service would have been impeded. These subaltern agents evinced a zeal, aptitude, and intelligence, rarely witnessed except in the French army. They were intrusted with the care of the visiting pass-books, the distribution of the food and medicines prescribed, the application of simple dressings, poultices, blisters, and similar services. They prepared with great skill the splints for fractures, and even applied, in a faultless manner, under the direction of the chiefs of the service, the dressings of amputated limbs.*

The happy results gained by the use of soldier-dressers in the Crimea, should not be lost. It may lead to the total discontinuance of assistant physicians, who discharge negligently duties they think of too little importance. Although they are not doctors, their embroidered collars make them appear like savans, and they are too often intrusted with medical responsibili-

* The following note upon these soldier-dressers is taken from a report of M. Lustreman, chief physician of the University Hospital, at Constantinople, made January 19, 1856: " Following the example of M. Thomas, at Gulhané, I have organized in this hospital a little company of soldier-dressers, and can affirm that the services rendered by these men have surpassed my expectations. They have enabled me to insure aid to all the wounded at a time when the energy of our medical force, whatever it may be, could not fully supply the want of numbers. Not only have I met in most of these a zeal worthy of all praise, but a tact and facility in learning their new duties, that I had not dared to count upon. One of these men, named Verdun, whom I made their chief, has twice, by a skilfully applied compress, arrested a hæmorrhage spouting from the femoral artery, thus giving me time to arrive and tie the vessel. It is obvious that the field of duty of these infirmary dressers should be quite limited, and strictly manual, and that they should never be allowed to meddle in the scientific, or even the artistic departments of the treatment. They are so many hands added to those of the chief of the service, and nothing more. This arrangement, extemporized in the army of the East, was of great assistance, and appears to me to have given a practical demonstration of the justice of the views long since expressed by Inspector Baudens, upon the subject of appointing special assistants in charge of the dressings."

ties.* Taken from the school-benches to serve in the armies, they lose the best time of life for study; their youthful years are spent in the camps, and when they return they feel neither the strength nor the courage to recommence their classical studies to acquire the doctor's degree. When the medical faculties finished by giving them a diploma, it was oftener on account of former services than their scientific attainments. Thus the health corps of the army has been filled with very inferior assistants.

The English, besides their regimental infirmaries, possessed four general field hospitals: one at Inkermann, two at Balaclava, and one at the Monastery of Saint George.† Their medical service, directed by the skilful and learned Sir John Hall, left nothing to be desired to

* The embroidered collars and sleeve-cuffs of the coat distinguish the *personnel* of the French service. They are crimson with the physicians, and light blue with the apothecaries. There is no difference among the several grades, excepting that the Major Physicians of the first class add a wand. The auxiliaries bearing commissions from the Minister of War have the uniforms of aides-major, but those employed by military intendants have no uniforms.—Tr.

† The hospitals were known as the General and Castle Hospitals of Balaclava, the General Hospital of the Third Division, and the General Hospital at Saint George.

The first of these was of stone, and had been used as a Russian military school. It consisted of two divisions, the front one containing large, lofty, airy wards, well isolated from the ground, and the other, of smaller apartments, built against the slope of a hill, with windows upon one side only. They formed two sides of a parallelogram on the east side of the harbor, the remaining sides being ranges of buildings used as stores and offices. Besides these, there were twelve or thirteen Portland huts in rows on the sloping ground above the harbor, which were used as wards for the sick. It proved to be on the site of an old graveyard, and unhealthy. It was best suited as a transit hospital, and latterly was almost entirely used for that purpose.

The Castle Hospital had a very fine natural position, occupying the whole of a long narrow ridge running nearly east and west, and joining Castle Rock with Marine Heights. It was three hundred and twenty feet above the level of the sea, and cut off from the adjacent higher ground by a deep ravine. The hospital consisted of a number of Portsmouth huts arranged side by side, with the ends facing the sea. Several other huts, on a model called the "Chester hut," were afterwards erected, until the whole ridge was occupied by thirty-one huts, all, or nearly all, used for hospitals or stores. The faults of

the end of the campaign. The attendants discharged their duties with zeal, under the impulse of active and intelligent female hospital assistants, at the head of whom was the celebrated Miss Nightingale. Beautiful, young, and wealthy, she sacrificed everything to the noble mission of alleviating suffering. This delicate young woman, mounted on horseback, might be seen passing from one hospital to another, looking after the sick of the three allied armies with a pious solicitude; and at the time of the typhus, she sent to the French and Sardinian field hospitals a large present of port wine and preserves of every kind.

The field hospitals of the English were extremely clean, which cannot be said of ours. The difference was in part due to the higher and more independent position of the English military surgeons, who exercise more authority in the enforcement of hygienic measures. Their allowance of food was better than ours. Tea, roast meat, and puddings held an important place,

ventilation that early existed were afterwards remedied, and the results showed an exceedingly favorable sanitary condition.

The General Hospital of the Third Division, in April, 1855, consisted of a number of Portsmouth huts, ranged in parallel lines behind the Third Division, and surrounded by the huts and tents of several regiments. The space was considerably raised above the general level of that part of the plateau, and had ample means of drainage. The soil was a tenacious clay, hardly capable of drainage, and liable to become extremely muddy at every rain. The huts were too crowded, their sides were banked with earth, and the ventilation was imperfect. These huts were gradually appropriated to other uses, and very few sick remained at the close of the year.

The General Hospital at Saint George was formed of a square of huts, similar to the Chester hut, upon a nearly level piece of ground not far from the top of the great ravine, and about five hundred feet above the level of the sea. The sides and roofs of the huts were double, and the ridge-pieces were raised so as to insure ventilation from between the sides. Each hut was fastened down at the angles by timbers, there was a porch at each end, and a row of swing windows of rough plate glass on either side. There was one large hut in this hospital made of corrugated iron, which, from its conducting power of heat, was found a bad material for the walls and roofs. In summer it became overheated, and in winter too cold, thus exposing the sick to injurious vicissitudes of temperature. The cases usually sent to this hospital were convalescents and ophthalmias from regiments in the front.—Tr.

and the physicians could order beer, wines of different kinds, rum, cognac, and whatever they judged proper. The extras must, however, be prescribed the night before. In the provision stores of these hospitals, I have even seen champagne, which they used to check vomiting.

The English soldier eats three meals in the hospitals, the regimen of a sick person being as follows:

For breakfast, at eight o'clock, he received 192 grammes (0.42 pounds) of bread, 500 grammes (a quart) of infusion of tea, made from four grammes (62·grains) of the leaf, and 24 grammes (three-fourths of an ounce) of sugar.

Dinner, at noon, was composed of a quart of broth, 500 grammes (1.1 pounds) of boiled or roasted meat, 128 grammes (four ounces) of bread, and 500 grammes (1.1 pounds) of potatoes, or other vegetables. Of beer, wine, or rum, they gave a variable quantity, according to the medical prescriptions.

For supper, at six o'clock, they served the same articles as at breakfast, and in the same quantities. Broth being given but once a day in the English hospitals, they have the advantage of being able to roast a part of the meat intended for the pot, a practice very advantageous to the sick.

The Sardinian field hospitals very much resembled our own, and the greater part of our regulations were adopted. The medical service with them is placed, as with us, under the authority of the military intendancy, and is not organized, as with the English, upon an independent basis. The learned physician-in-chief, M. Comizetti, was well seconded by able and experienced practitioners. Their field hospitals were located upon the high plateaux of Kamara, upon the Cape of Balaclava, and were each composed of 42 fine barracks, of average capacity for 36 beds, planked and well kept. The beds were made of two wooden trestles, supporting three boards, upon which were placed a mattress, a pillow, a pair of sheets, and two blankets. The officers had, in addition, a straw-bed, a bedside table, and a mat spread down by the side of the bed. Their regimental infirmaries were arranged like ours, and altogether they counted 1600 beds, a high number for an army of 15,000 or 18,000 men. Never more than 1200 of their beds were occupied at one time. The Piedmontese army was heavily afflicted with scurvy, but the typhus touched them but lightly. In each section a Sister of Charity presided over the distribution of food and medicines, supervised the attendance of the sick, and directed the attendants. In the kitchen, the dispensary, the apothecary-room, the laundry, and the wash-room, was always found a devoted and intelligent Sister. They went daily to the market at Balaclava, to purchase provisions; and their thoughtful charity had provided the field hospitals with a poultry-yard of five hundred hens, which were fed from the refuse of the table. The salary of these Sisters was five hundred francs, and they received, besides, two daily rations of camp provisions. They filled very nearly the places of our infirmary-majors. The Sardinian physicians were also aided by soldiers, in bleeding, and they also took charge of the prescriptions and visiting pass-books. To each field hospital was attached a skilful knife-grinder—an excellent measure, which might be imitated.

I visited the Russian field hospital at Belbec, which was well arranged, and provided with good furniture. They had double beds, with separate mattresses, sheets, and blankets, thus economising space, but not without danger to the sick. More than half a century ago, this unhealthy practice was discontinued in France. The barracks, arranged for 120 places, in four ranges of beds, were well built, but not ventilated. In the finest weather the doors and windows were closely shut, the air was heavy and offensive, and typhus made extensive ravages. My secretary, M. Benjamin Crombez, was attacked by it, after remaining only a single hour in this field hospital. It is remarkable that the memory of past suffering should not prove more instructive, and that the most terrible lessons should be lost. In 1829, their army of the Danube, attacked by typhus and the plague, lost 60,000 men. The number of men that recrossed the Pruth did not exceed 10,000 or 15,000 combatants.* Like the English and Sardinians, the Russians employed women for attendants. To their Sisters of Charity, who were much esteemed, were added the lady attendants, for the most part composed of officers' widows, who brought voluntarily to the Crimea the tribute of their pious devotion. They were employed in the laundry, the kitchen, and the apothecary-room, took care of the wards, and watched over the fever patients and the wounded, with indefatigable zeal. Upon some of the beds we saw dead soldiers, with their faces uncovered; little wax candles burned at the head of the bed; doubtless some religious rite; the adjoining sick took no notice of it.

Two Russian physicians, taken prisoners with their attendants in the attack on the white works at Sebastopol, were brought to the general's quarters. One of them, wounded in the head, was taken care of in the field hospital, while the other, who was unhurt, proved a very skilful surgeon, and was placed in special charge over the wounded Russians. Like the rest of his countrymen, he practised amputation by the circular method, cut the integuments behind to facilitate the flow of pus, and kept together the lips of the wound by a wick of waxed lint, after the manner of Sédillot, of Strasbourg —a practice which gave good results. The Russian physicians spoke French, and lived with our physicians; they were afterwards exchanged for some of our men. Their assistants were so skilful in ligating arteries and in applying dressings, that not a single secondary hæmorrhage occurred.

Felchers replaced our soldier-dressers in the Russian army. They rank as sergeants, are a permanent class, and are usually recruited from among the orphans of soldiers dying in the service of the empire. The government brings them up, and gives them a crude education in surgery and military hygiene. In the hospitals, we found one felcher to every seventy-five sick; the regiments are composed of four battalions of 1000 men each, and they have one to a battalion. The sur-

* "Campagnes des Russes dans la Turquie d'Europe, en 1828 et 1829," by Colonel Baron de Moltke.

geons under whose orders the felchers are placed, praised very highly their behavior, intelligence, and subordination.

As to the physicians of the Russian army, they are either the orphans of officers, brought up and educated at the expense of the government, or physicians taken from the ranks of the profession, and graduated from their great universities. The state requires of the former six years' active service; the latter can resign. After having served twenty-five years, they may both claim a retiring pension, which equals half their pay according to rank. After thirty years' service the pension is the same as if in active service. Every five years after the first six, they receive an increase of pay. Like the officers of the army, their grades begin with the rank of captain, and rise to that of general, sharing with the other officers all military honors, as well as dangers and privations. The Russian soldiers carry in their knapsacks a compress and roll of bandage—precious resources, which allow of the instant and efficient application of a first dressing upon the battle-field, where articles for dressings are so often wanting.

CHAPTER II.

SURGICAL OPERATIONS.

Most of the surgical operations are made in the field hospitals, and consist of extraction of balls, amputations, and resections.

In the Crimean campaign, the severity of the wounds was not only due to bullets and shell, so terrible in a siege, but also to the employment of arms of precision, and the substitution of conical balls for round. These improvements were known to the Russians, and were, before the walls of Sebastopol, as murderous to the besiegers as to the besieged. But still, thanks to the genius of our nation, to the intelligence, activity, and hardihood of our soldiers—these arms "essentially French," as Marshal Bugeaud says, were peculiarly destructive in their hands.

The arrow had for some time suggested the idea of giving to spherical projectiles an elongated form, in order to diminish the resistance of the air.

It has been long attempted to substitute for the smooth bore, tubes grooved in spiral form, which should give to the ball a rotary motion; this idea is the sequel of the preceding one. This rotative motion can only be gained by forcing the ball into the groove, and to remedy the inconvenience of driving it in with the blows of a mallet, the *flattening* method has been proposed, especially by M. Delvigne. This is done by driving the leaden ball smartly with the ramrod, against a little projection in the chamber of the gun. In the forcing by *dilatation* afterwards proposed, the cylindro-conical balls are hollowed at their base, and the gases resulting from the ignition of the powder, act upon this concave surface by a violent expansive effort, forcing the lead into the furrows of the barrel. The former system has been in use in France more than twenty years, but the second only a few, and has likewise been adopted in England. Experience having shown, that after passing a certain distance, the longer axis of the projectile tends to incline from the line of flight, this has been remedied by the ingenious and simple use of circular grooves, made around the ball near its base, which perform the office of quills, or plumes, which, in the flight of an arrow, are so important.*

Arms of precision have received new improvements, known by the names of their inventors, Poncharra, Minié, and Thouvenin, and to the latter belongs the honor, of having produced the carbine called *à tige*, so much esteemed among the battalions of foot chasseurs. The *tige* is a little cylinder of steel, screwed into the middle

* The Russian rifle was little inferior to the minié in range or force, while its conical deep cupped ball was heavier. The severity of the primary action on the part struck, and the aggravated evils which followed their wounds, combined to exercise a most prejudicial influence on the surgery of the war.—TR.

of the breech, and of less diameter than the barrel. The powder lies around the *tige*, and the expansion of the ball is effected by flattening it against the top.* The range of these balls is more than 1,400 yards, but beyond 800 yards it is not exact. Experiments have shown, that a round ball, at a range of 600 yards, will not pass through two-inch fir planks, while a cylindro-conical ball, thrown from this carbine, will easily perforate eight such planks.

Wounds present different characters, according to the velocity and shape of the projectile. The hole where a spherical ball has entered, is round, depressed, and smaller than that where it comes out. The path which it traces through the limb, is conical, and goes on enlarging, and if it meets the sheath of a tendon, or a bony surface, it is often much deflected from its course, so that it is not rare to find the place of outlet, which is more irregular and bruised, with the lips raised and jutting not directly opposite the entrance.†

The place of entrance of a cylindro-conical ball is oval, and sometimes linear, as if made with the point of

* This arm was furnished to the Chasseurs de Vincennes. The ball weighed 1 ounce 5 drachms and 16 grains.—Tr.

† It is not always easy to distinguish the gunshot wound of entrance from that of exit. According to Dr. Macleod the former is more regular, and less discolored than the latter. The inversion of the lips of the wound in one case, and their eversion in the other, is not always clearly marked, and if the velocity of the ball has been great, and no bone has been struck, there is very little difference in either the size or discoloration of the wounds. The enlargement of the wound, as the projectile enters, is common when the velocity is nearly spent, or when a bone or aponeurosis has checked the force, and changed the direction of the ball. This is especially true of the conical projectiles. Muscular wounds from gunshot, almost always heal by suppuration and granulation, the exceptions being extremely rare. The opening of a wound may be apparently smaller than the projectile. At Scutari, a piece of shell weighing nearly three pounds, was extracted from the hip of a man wounded at Alma, which had been overlooked two months, and entered by a small aperture. Larrey gives a case of a ball weighing five pounds, which was extracted from the thigh of a soldier. It had not been detected by the surgeon making the dressing, and incommoded only from the weight. Fragments of *nine* and *twelve* pounds weight have been reported as buried in the thigh with no suspicion for a time of their presence. The elasticity of the soft parts enables us to explain the smallness of the entrance wounds in these cases.—Tr.

a sabre; the ball which General Thomas received in the groin at Alma made a wound exactly like that of a sword or bayonet. Their line of passage does not appear to take the form of an elongated cone, which is probably owing to its twisting motion, and deviates much less from its course, than the round ball. Neither the sheaths of muscles, nor tendons, nor bones resist the penetrating points of conical projectiles, the lead is perhaps battered or turned, so as to present its greatest surface, which is about a third of an inch in area. In this case the outlet, which does not usually differ from the entrance wound, except that it is larger, and with its lips more bruised and projecting, presents unusual dimensions with rags of flesh very irregularly torn.

If a round ball meets at an inclined angle, a curved bone, as the cranium or a rib, or a long round bone as the femur, it often passes around, instead of breaking it, but a conical shot almost always breaks it into splinters. We therefore observed in the Crimea a proportionally larger number of fractures from balls than in our African wars. It should nevertheless be remembered, that a round ball scarcely weighs an ounce, while the others are nearly twice as heavy. When a conical ball strikes, it may perhaps be inclined, so as to enter crosswise. In such cases, which are rare, the entrance and path are very large; we usually find the lead at no great depth. Its extraction is easy.

In 1830, when I joined the army sent to Algeria, it was the rule and practice to lay open with a bistoury the holes of entrance and exit made by balls. The most approved teachers advised large incisions of the skin and subjacent tissues, to favor granulations in the injured parts, and prevent their choking up, thus avoiding the evils which ensue from such cases, as for example gangrene. This bloody operation, called *unloosening*, was much more painful than the wound made by the ball, but no one questioned its efficacy. It had become, so to speak, a medical dogma. In the early conflicts in Africa, at Sidi-Ferruck and Staoueli, I saw with astonishment a large number of wounds, which, for want of time, had not been enlarged with a cutting instrument, healed without mishap, and even more quickly than those through which the bistoury had passed. In the Crimea, I observed with satisfaction, that the unloosening of wounds had not a single partisan. Although it still finds academic partisans, it has been styled a useless and barbarous practice. In these terms I spoke of it in a work published in 1836,* and nothing has occurred to lead me to change my opinion. I have even proved, that free openings do not prevent accidents, when the wound contains foreign bodies, as wadding, pieces of clothing, or lint, drawn in by the projectile, or when the ball itself, whether entire or broken, has come in contact with bone, breaking it into splinters.

Sometimes these splinters of bone remain in the flesh, irritating it like thorns, in which case the best remedy is, to extract these foreign bodies. Should we trust to the eliminating agency of suppuration, as is still advised, in getting rid of these splinters? Failures in this are so common, that it is evidently preferable to take them all out as soon as possible, whether adherent or not, with the view of simplifying the wound. A simple wound heals regularly, without inducing a variety of painful complications, which every moment endanger the lives of the wounded. Lisfranc said, we should wage a *partisan warfare* with these complications, that is that we should combat them bistoury in hand. It would be better still, to prevent them, by removing the splinters at once. Their retention leads to endless suppurations, with constant suffering, which is increased whenever a piece of bone works out, thus wasting away the vital forces, causing marasmus, reabsorption of pus, colliquative diarrhœa, and death.

On the contrary, when the wound contains not a bony splinter, but a round smooth ball, of which the surgeon cannot readily find the track, it is wiser not to multiply his searches, and to spare sufferings to his

* *Clinique des Plaies d'Armes à feu,* 1 vol. 8vo. Paris. 1836. J. B. Baillière.

patient, as its presence is less irritating than the angles of a splinter, and furthermore, in the course of healing, the ball will work towards the surface, and will be more easy to reach.

The extraction of balls is often a difficult operation, because they traverse tissues of various density, and therefore of different degrees of resistance, which cause them to deviate from a direct line. A ball striking a rib obliquely, does not always penetrate the chest, but passes along the curve of the bone, upon which it is held by the elasticity of the skin, neutralising its centrifugal force. When a ball penetrates the tissues, it tears them by pushing forward, like a blunt point; and at the end of its course, it often meets in the skin a resistance which it cannot overcome, and it remains just beneath the surface. In this case, by taking it between the thumb and fore finger of the left hand, and cutting the skin over it, with a little gentle pressure down and behind, it will roll out. If it resists, it is needless to enlarge the incision, and we must ascertain what holds it. The obstacle may be due to a thin and transparent layer of cellular tissue, with which the projectile has enveloped itself in its thrusting course at the end of its flight. It forms a little sac, which only needs to be opened to allow the ball to escape.

I can demonstrate this fact conclusively, and have called this the *primitive sac*, to distinguish it from the *definitive sac*, organized around balls when left to themselves, and allowed to remain, thus becoming naturalised in the tissues.* These singular guests may ever after remain quite harmlessly in their sacs, or at times, the pressure caused by the weight of the lead may irritate and soften the sac; the ball opens for itself a path step by step, the void behind closes gradually by granulation, a process of cicatrization

---

* This observation has not been made by other writers upon Military Surgery, and the cellular envelope, although occasionally found, is rare.

Muscles wounded by gun-shot are apt to shorten during the cure, unless care is taken to prevent it. The result is always unfortunate. —Tr.

which presses it forward upon its journey. The progress is slow and almost imperceptible, so that only in the course of years will a ball which was in the groin, descend to the heel.

Wounds made by fire-arms, being essentially contused, cause a strong inflammatory reaction, which, by a variety of accidents, may lead to gangrene, and demand oftentimes an energetic treatment. In these wounds, the best therapeutic agent I have found to be cold, produced by ice. The illustrious Percy used cold water, in treating gunshot wounds. I have followed his example with results so favorable, that I was led to study carefully the action of refrigerants. When the inflammation exceeds certain limits, cold water is not sufficient, and we must then resort to ice, either alone, or mixed with sea-salt, to increase the intensity of the cold, which should be regulated according to the violence of the traumatic inflammation. Ice should never be in direct contact with the surface. We begin by placing upon the inflamed part a simple compress of linen, soaked from time to time in cold water, and then put pieces of ice in the folds. If the refrigeration does not procure more than a moderate relief without destroying the deep and painful sense of heat, or, as the patients express it, if the ice seem to *burn*, the cold should be increased, by the addition of salt. Nothing is easier than to avoid the abuse of refrigerants, and the accidents resulting from their use. The contact of cold upon an inflamed part causes an agreeable and decided sense of relief, and this is an infallible guide, which should be carefully observed. The refrigerants should be continued, so long as they produce this sensation, and should be gradually withdrawn when they cause a disagreeable sensation of cold dampness. This does not occur until the fire of inflammation is extinct; and if continued, the ice would cause the most severe pain, from the abstraction of the normal amount of caloric. The patient is therefore the best judge to consult. Before applying ice, the physician should ascertain the general condition of the patient. If the constitution has been enfeebled by fatigues and privations, if he fears that the broken vital energies will fail, and that a salutary reaction will not ensue, he will give stimulant drinks instead of refrigerants, and place over the wound a thick layer of lint, to retain the heat. The use of ice, in such a case, would be a monstrous absurdity.

The opponents of the refrigerating treatment apprehend gangrene, or at least the reaction and check of perspirations. It is difficult to understand *à priori* how a limb can without danger be covered with ice several days at a time, when a single icicle held a few moments in the fingers, leads to a beginning of congelation, keen pain, and a feeling of insupportable constriction. It is because in the one case the ice acts upon an inflamed surface, and in the others upon a healthy part. Inflammation imparts to the region of which it is the seat, a remarkable power of resisting cold. Hunter, after having frozen the ear of a rabbit, by immersing it in a dish of pounded ice, could not freeze it again, when inflammation had set in. This was a great discovery. The organic, normal, or physiological heat, existing in health, should be distinguished from the abnormal heat of inflammation. The former is indispensable in regulating the animal functions, and cannot be withdrawn without peril; we know that a simple chill may be dangerous. The caloric produced from inflammation, if it does not exceed a certain degree necessary for healthy action, ought not to be lessened. It is when it appears in excess that it is full of danger, and excites a host of disastrous results. It is better in these cases to resort to refrigerant applications, than to local or general bleeding. Cold is a sedative, calming the pain, and preventing the determination of blood to the affected part, while bleeding, by suction, and the painful pricking, attracts the blood, and congests the wound. Cold acts upon the patient as a tonic, while bleeding weakens; cold is a most energetic agent for arresting inflammation and preventing it from spreading, while bleeding is often powerless; and cold tends to localize the phlegmasia, imprison it in the wound, and prevents its sympathetic transmission to the great viscera, especially the heart, the reaction on which excites fever. The inflammation has sometimes been so intense, that I have had to apply for several days, to complicated wounds that were choked up, a freezing mixture of —14° Centigrade (—4° Fahrenheit).

During the insurrection of 1848, I kept ice during forty days upon the leg of a wounded officer. A quarter of the substance of the tibia, ground to pieces by the projectile, had been extracted to simplify the wound; an amputation was avoided, and fifteen months after, this officer laid aside his crutches, and could walk freely. This was one of the finest triumphs of conservative surgery.

Treatment by ice may not only be applied to wounds received in war, but also to lesions resulting from accidents, as dislocations, contusions, and fractures, and especially to strangulated hernia, which it relieves without an operation, with great success.* It is proper to limit its use to lesions *caused by violence*, because the inflammatory action is then decided, simple, and free from predisposing individual influences. Used on a multitude of patients at Val-de-Grâce, where I directed the surgical service during ten years, the treatment by ice has been tested; military surgery has adopted it and finds it answer.

When a splinter of a shell strikes by one of its angles, it makes a large cut, very clean in appearance, but one which cannot heal until after the adjacent parts, that have been killed by the violence of the shock, are eliminated. These wounds had in the Crimea a remarkable tendency to the characteristics of hospital gangrene.

When cannon balls nearly spent are rolling upon the ground, they must be carefully avoided, even when the motion is very slow. A grenadier of the guard, lying upon his side on the ground, was suddenly killed by a ball which broke his spinal column. This ball had so little force left, that by a singular coincidence it lodged in the hood of the soldier, and was found there.

* On the 29th of May, 1854, I read before the Academy of Sciences a memoir upon the value of ice in the reduction of strangulated hernias.

Bombs always produce very serious wounds, and make in the chest and abdomen such terrible breaches, that art is powerless for relief. When a ball, or a piece of a bomb, has taken off a limb, the wounded man very often evinces the effects of a general shock, and, as soon as the stupor begins to pass off, the hand of the surgeon is necessary, to trim the wound. If left to the unaided efforts of nature, the violently torn tatters, mixed with tendinous portions of unequal length, and bruised pieces of bone, almost invariably tend to a mortal gangrene. The rupture of arteries may cause hæmorrhages that prove quickly fatal. General R—— died in a few moments from a hæmorrhage caused by a musket ball, which divided the popliteal artery. His life might have been saved by compressing the vessel until a surgeon arrived. There are some cases in which the violence of the blow itself brings with it an effectual remedy; the tearing off of a limb leads to a retraction of the arterial tissues, and the opening of the tube closes upon itself, thus opposing a barrier to the expulsion of the blood.

Sometimes the effects of bombs are horribly fantastic. General Pecqueux de Lavarande was literally cut in two by a bomb which exploded between his legs, the head remaining on one side with an arm and a leg. Other effects more singular, but not as horrible, were many times observed during the siege of Sebastopol. We know that bombs, in traversing their parabolic curve, give out a peculiar hissing sound, which enables one to dodge them, and by dropping flat upon the ground, to avoid the explosion. It has so happened, that at the instant when the soldiers are stooping to drop on the ground, the bomb in its path followed the curvature of the spinal column, and crushed it the whole length, causing instant death. At another time, when the shock was less violent, the spinal column resisted, the skin from its elasticity was not broken, and the subcutaneous bloodvessels were alone torn. In this instance, the blood accumulated near the os sacrum, at its lowest part, and the sac was punctured by the subcutaneous method of M. G. Guérin, to avoid danger from the introduction of air into the closed cavities. A black grumous blood escaped, and the patient recovered.

At another time, we found on the battle-field a dead body which showed no external marks of wound, and whose death they ascribed to the *wind of the ball*. This is now acknowledged to be an error, for we have seen a ball take off a knapsack from a soldier's shoulders, his cap, and even his pipe, without leaving any effect of its passage; and, again, the parts under the skin have often been found bruised to pulp, and the bones broken to atoms. The wind of a ball could not do such damages, but it was the ball itself, especially in some of its last rebounds, towards the end of its flight. The elasticity of the skin explains why it can remain unbroken, although a ball may come in contact with the body.

At the battle of the Alma, the splinter of a shell struck General Canrobert on the breast, bruising the pectoral muscle, while it scarcely broke the skin. General Bosquet received on the after part of his chest just below the shoulder blade, a splinter of a shell. The skin became ecchymosed, but was not broken, although three ribs were fractured on their inner side, in a curved form; they were replaced with difficulty, and left a very apparent depression, which was quite perceptible to the touch. This fracture, probably complicated with a rupture of the lung, admitted into the chest an effusion of blood which crowded the pulmonary tissue against the spinal column, and hindered the introduction of air into the bronchial cells. The general was very skilfully treated by M. Secourgeon, chief physician of the 3d Corps, and by Dr. Combarieu, who had him carried from the trenches to the Lancaster batteries, where he was bled. M. Combarieu then took him to Pau, to the house of his aged mother.

The department of our science which treats of amputation is of first importance. When ought we to practise it, and when avoid it? Upon these two questions the Crimean war has shed a flood of precious light. During the wars of the empire, no matter in what part of the leg the wound was situated, were it even the heel, they practised amputation at about four or five fingers' breadth below the knee. The stump, disguised by a long pantaloon, was fastened to *a wooden leg* of the simplest kind and not liable to get out of order.

There was here an advantage which many surgeons do not like to give up. During the last twenty years however distinguished practitioners have laid down the precept, that the amputation should be made as low as possible, and always just above the ancles, that is, the ankle bones. At this point the leg is smaller than above, the wounded surface is less, there is less inflammation, ulterior accidents are reduced, and cases of healing are numerous. This rule involves the general principle, that we should always amputate as far from the trunk as possible. The office of the surgeon is, above all, to save the life of the wounded, and his conscience bids him prefer the safest way. Other reasons operate in favor of the sub-malleolar amputation. The important articulation of the knee is preserved, and by the aid of an artificial limb it is easy to conceal the mutilation. This consideration is not one of indifference to a young man who has his way to make, nor even to a superior officer, commanding on horseback. A lieutenant-colonel upon whom I performed the operation was able to continue in the service, and is now a colonel. It is true that the apparatus furnished to the soldier who has suffered an amputation above the ancle bones costs two or three times as much as the classic peg, but the State will never be run in debt by so small an affair. I demonstrated in 1839,[*] that we may take off the whole foot without having recourse to amputation of the leg, taking a flap from the soft parts to cover the wound on the instep, or better still, on the heel.[†] In the latter case the amputated person walked well without the aid of any apparatus, by bearing the weight of the body upon a short wooden leg fixed to a bootee with a high heel. I have met some amputations performed in the middle of the calf of the leg, and I have blamed them

[*] New method of amputations, first memoir, *Tibio-Tarsal Amputation.*—Paris, Germer-Baillière, 1842.

[†] *Nouveau procédé de l'amputation tibio-tarsienne,* with a Plate.— *Gazette des Hôpitaux,* Feb. 1850.

exceedingly. Soldiers should never be subjected to such experiments, and the army council of health are quite right in maintaining the wise and traditional rule, which prohibits military surgeons from employing modes of treatment and operations not sanctioned by experience. An amputation in the middle of the calf has serious inconveniences. The volume of the leg, much greater at this place, gives a large wound and increases the chances of mortality. The stump is left too long, so that a wooden leg cannot be readily applied.

A truth, which numerous facts observed in the Crimea now allow me to affirm, is—that when it is not possible to amputate the leg, the disarticulation of the knee is to be preferred to the amputation of the thigh. The former succeeds oftener than the latter. The disarticulation of the knee should be done immediately; and it is a point acquired henceforth for science, that we hazard success by delay. In fact the volume of bone, even in a state of health, is not in harmony with that of the soft parts, and the disproportion will become greater as the patient loses flesh by prolonged suffering and profuse suppuration. This amputation preserves to the patient the free use of the hip-joint, and gives a solid point of attachment for an artificial limb.

For half a century surgeons have discussed the question, whether it is better to perform an amputation as soon as possible, or to wait some days, if not weeks. At the present day the first of these two opinions has decidedly the preference, and the results of observations in the East militate in its favor. We shall observe, for example, in a review of immediate and delayed amputations practised in the hospital at Gulhané,* that three immediate tibio-femoral amputations succeeded, and that five consecutive amputations performed after some delay at Constantinople, failed.

Before the Crimean war it was a generally received rule, that a femoral fracture by fire-arms rendered an amputation necessary. There is reason to believe that, thanks to my new apparatus for fractures, this rule may

* See Appendix.

be considered too absolute.* These appliances have the advantage of keeping the limb in its normal position without compressing or wasting it, and of keeping the fracture in perfect rest by permanent extension, counter-extension, and adjustment, worked by the aid of elastic straps replacing perfectly the contractile action of the fingers. Inflammation is sooner allayed, the thigh almost completely naked is exposed to the salutary influences of the air and light, and the surgeon can constantly follow with the eye the progress of the case, dispense with assistants, and make applications and dressings to the wounds very easily. In the Crimea and at Constantinople, in the experience of our most skilful physicians, Lustreman, Thomas, Salleron, Maupin, and Marmy, many legs and thighs were saved by using this apparatus for

* This apparatus is thus described in the *Comtes Rendus* of the Academy, 1853, first part, p. 854.

It consists, first, of a kind of box open on the top; second, of an inclined plane; third, of three cushions of hair; and fourth, of two graduated compresses with elastic strings for fastening.

The open box for lodging the affected limb should be about thirty inches long, eight wide at the place for the thigh, and only seven where the leg and foot are placed. It is made of four pieces; the bottom, two sides, and a foot board, the latter moving upon hinges and capable of being fastened by a hook. The bed-piece is horizontal, the sides about nine inches high and pierced with three rows of holes placed one above the other about as large as to admit the finger. The foot-board is of the same height or a little higher. The inclined plane is designed to receive the broken limb, and the cushions of hair to form a bed on the bottom of the box. The two graduated compresses should be of the length and thickness of the fore-finger, and are to be placed across the limbs above the fractured parts to assist the action of the adhesive straps. The limb is placed upon cushions, the parts adjusted and held in position by elastic straps, fastened by adhesive straps operating by extension and counter extension in such a manner as to supply a pressure, much like that of the fingers of the surgeon. The numerous holes afford a choice of selection for points of attachment of the elastic straps, and enable the surgeon to apply his tension as he may have occasion. The foot-board has also holes for a like purpose. The heel is supported by a little pad under the tendon of Achilles, and in a later description the sides of the box were fastened by hinges to the bottom so as to adapt their inclination to a given case.

Descriptions of the apparatus and of cases treated under it, were read by the author before the Academy of Science at Paris, June 30, 1851, May 15, 1853, Aug. 7, 1854, and Jan. 15, 1855.—TR.

fractures. It is first necessary to extract all splinters of bone; their presence in the flesh leads to an interminable suppuration and almost always to death. After these are extracted, the wound is given such an inclination as to allow the ready escape of pus, and nature is left to work the cure without hindrance. Many cures have been obtained without leaving noticeable deformities. The unsightly callosity, or place of joining of a badly united fracture, with deformity of limbs, has been successfully reduced by the use of this apparatus, even after several months.* The shortening of the femur corresponding with its loss of bone may generally be disguised by using a high-heeled boot.

Amputations of the thigh are the more serious on account of their proximity to the trunk, it is therefore very important to avoid them. Although disarticulation at the knee ought to be done immediately, that of the hip does not succeed so well, but on the contrary, should be delayed some days after the injury is received. This remark is very important, for it allows of attempts being made to save the limb. The upper extremity of the femur is chiefly formed of spongy tissue, more easily traversed than solid bone, and the ball finding less resistance, does less injury. We may then apply the apparatus, for we run no risks, as, if it fails, we may still resort to amputation.

In the upper extremities, we may very often avoid amputations, and preserve the limbs, not only by taking away the splinters, but by resection, a method of operating, which gives most admirable results. These resections I had often practised on the battle-field, or advised and explained in books and lectures. It was with much satisfaction that I saw the surgeons in the Crimea more cautious as to amputation, and performing resections as often as possible, instead of taking off the entire arm. Resections apply to the salient angles of fractures, happening in the body of long bones, or at their articular extremities. The periosteum, or membrane covering the bone, must be pre-

---

* See Appendix.

served with the greatest care; as M. Flourens has shown, that it is this membrane that secretes bony tissues, and that it will reproduce them, if left in place. The grand triumph of resection is when it is practised upon the head of the humerus. An officer of high rank, M. Bertier, who submitted to this operation, is now colonel of the 86th regiment, and uses very well the arm that was operated upon, although it is a little shorter than the other. A sergeant-major, M. Plombin, upon whom, at the age of twenty-three, I performed the same operation in Algeria, is now colonel of the 1st regiment.*

Thanks to resection, an isolated fracture of the radius, or of the ulna, does not necessarily require the loss of the limb, and I have, with success, taken out almost the whole of one of these bones. Even where both bones are fractured, they are not always, unless with grave complications, a subject of amputation. We may say as much of fractures of the body of the humerus. Resections are useful, not only for saving the limb, but for securing a more ready cure. We should observe as well as possible, our rules of conservative surgery. Especially in fractures of the hand, it is important that we should be imbued with this precept, and that we should apply it in all its vigor. We should never forget that the misshapen stump of a finger may still be very useful. Eight years ago, in June, 1848, they brought in a captain, to have his right wrist amputated, in consequence of a gunshot wound. I succeeded not only in avoiding amputation, but in saving the little finger, half of the index finger, and the thumb. This officer can still hold a sabre, and has con-

* The author, in a paper read before the *Academy of Sciences*, vol. xxvi., 1855, performed fourteen operations for resection of the head of the humerus, all of which, but one, had succeeded. When the head of this bone has been broken by a ball, he regards resection as the rule to be followed, and amputation the exception. By his method, the limb did not necessarily remain suspended from the shoulder, as sometimes happens, but he obtained in most cases a good joint, where the glenoid cavity of the scapula was intact. The end of the amputated bone should be brought in contact with this cavity, and all muscular fibres should be carefully removed, only a simple incision was made and no flap.—TR.

tinued to serve. I met him in the Crimea, the colonel of a regiment, and to bring himself to my recollection, he showed me his hand.

We cannot so often practise resection upon the lower extremities, especially in time of war, when the wounded are liable to long and painful removals. Being the means of support, the lower limbs must be more solid than the arms. The very voluminous muscles render less accessible to the hand of the surgeon, the splinters of the femur than those of the humerus. A comminuted fracture of the two bones of the leg, is often a cause of amputation; but if the wounded can endure the perils of transportation to a well furnished establishment, we should try to save it. When the tibia, or the fibula, is only broken, the resection, or even the simple extraction of the splinters, often suffices to lead to a cure. Perforations of the foot, by bullets, are less grave than were formerly supposed, and by the extraction of the splinters, we may almost always avoid amputation. In June, 1848, M. Thayer, now a senator, received a wound of that kind, and the extraction of the splinters with long continued application of refrigerants led to a perfect cure.

The most terrible enemy which the surgeons of the army of the East had to encounter, was *Hospital Gangrene*. This pestilence, like typhus, springs from concentrated and prolonged mephitism, so difficult to avoid in armies stationed for a length of time in crowded tents. It arises spontaneously, is propagated by the air, or by means of direct contagion by the deposit upon a healthy wound, of matter coming from a surface infected with gangrene. The second mode of extension would not be so serious, if care was taken to burn all the linen which had been used in dressing gangrenous wounds, with the view of preventing its being used again. The air is so obviously the vehicle of infectious miasms, that hospital gangrene always shows a tendency to increase or decline, according as the wards are more or less encumbered.

Hospital gangrene attacks the wounded, as well as those whose wounds are not entirely cicatrized; and at the time when the poor wounded men have just reached the end of their cure, and are preparing to return to their families, they perish, victims of this horrible disease! We recognise this terrible malady, when the wound grows dry, and becomes painful and slate-colored, with black patches. A gangrenous disorganization breaks out, attacking by preference the cellular tissues, of which it eats out deep excavations. At times, instead of a humid gangrene, the skin becomes covered with a dry scab; and at other times, while the wound is healing on one side, it is increased in size by ulcerations on the other. This is called, "ulcerated hospital gangrene." At the beginning, a reddish violet circle from a fifth to a third of an inch wide, is formed around the circumference of the wound. In three or four days, this falls into a gangrene, and another circle succeeds; it becomes gangrenous in its turn, causing great destruction of the substance, often attended with alarming hemorrhages. This kind of local typhus does not hesitate to invade the whole organism, and death presently follows, if art proves powerless, or cannot seasonably intervene.

The first remedy, and that without which all others are powerless, is to isolate the infected, in non-contaminated places. This isolation is at the same time demanded—in the interests of the sick, tainted with the pestilence, and in those of the neighboring sick, who may be exposed to contagion. The tent is here an excellent resource, especially if each one of those attacked can have one by himself. The air is then easily renewed, as it is only necessary to fasten up the sides a few inches from the ground, to secure a constant and very salutary ventilation; the bed being above the openings, the patient does not suffer any inconvenience. When hospital gangrene is once established in a hospital, it is very difficult to get rid of its contagious miasms; and it is necessary to abandon it for a time, whitewash the walls, frequently sprinkle the floors with chlorides, and fumigate thoroughly. In this way alone can it be conquered.

The local treatment consists in cauterizing with a red hot iron, or with the perchloride of iron, a powerful caustic, which penetrates easily into all the sinuosities of the wound; this treatment M. Salleron employed with success. Dressings of lemon juice, pulverized charcoal, and quinquina, with camphor, are secondarily useful with caustics. Continued lotions of cold water, falling drop by drop, are also a disinfectant, an excellent modifier of the disease, and a constant sedative of pain. Injections of the tincture of iodine, which M. Velpeau commonly employed, also give advantageous results. But still, these are only auxiliaries; caustics and cauteries alone can stop the progress of hospital gangrene, and suppress the sources of the putrid fluids that infect the whole economy.

There were 5,000 to 6,000 cases of frost-bite, during the two winters of 1855 and 1856; and the facts observed, presented noticeable peculiarities. In 1855, the cold was not very intense, but there were abundant rains, and the soil remained soaked a long time, especially in the trenches. The feet of the soldiers, immersed in the icy water, suffered from the effects of frost, similar to that in 1836, before the walls of Constantinople. It produced a tumefaction, attended with redness and gangrenous spots, more or less defined. Freezing reproduced the six degrees of change, admitted by Dupuytren in burns; we know that cold and heat produce the same effects. The disorganization acts by a humid gangrene.

On the contrary, in the rigorous winter of 1856, when the thermometer often descended to five degrees below zero of Fahrenheit, we often observed dry and sudden gangrene. The sick, when being transported to the field hospitals upon chairs, and the soldiers sleeping in tents, had their extremities frozen. The extreme cold expelled the liquids, the feet dried up, were reduced in volume, turned, so to speak, to parchment, became a dirty white, and finally, formed a dry, black eschar, like in a mummy.

In Russia, when they travel in sledges without taking the precaution of covering the nose and ears with furs, it happens that the parts in like manner become suddenly white, wrinkled, and deprived of life. The ears of our soldiers were preserved by red cloth caps called the *chachia*, which were distributed to them. Among the soldiers who conducted the wagon trains, and who were obliged to be out in all kind of weathers, we most frequently observed the effects of rigorous cold; many of them had one or more fingers of the rein-hand frozen. Frozen feet were very common in all the regiments. The soldiers took care to be always moving, and would call out to those of their comrades whom they saw standing still, careless or indifferent, " you will freeze;" unfortunately, they were not mistaken. When cold acts upon the whole surface of the body, it does not produce very much suffering, but a simple dulness throughout, and an irresistible desire to sleep, which is fatal if indulged in, especially after having drunk alcoholic liquors in excess.

If limited to the feet or hands, the action of cold is first announced by a painful numbness. A horse-soldier whose right hand had been frozen, said that he had first felt a numbness in the ends of his fingers, but that soon after they seemed to warm, and he felt it no more. The recurrence of cold rendered it difficult for him to hold the reins, and soon his fingers formed a rigid hook around which he fixed them. He was so imprudent as to approach a fire, and lost all his fingers by gangrene. He should have begun by rubbing his hand with snow, and should only have recalled warmth and life by degrees.

Until the parts struck dead by congelation drop off, the suffering is very slight; the appetite remains, and there is but a very moderate amount of fever of elimination. They were satisfied, under my advice, to wrap the limb in soft folds of wadding, the light and silky contact of which is agreeable; the patients enjoyed their food, and often congratulated themselves that the affection was not serious; but unhappily, when, a little after, the toes, or a half of the foot, or both feet, beginning to putrify, separated from the body, leaving exposed a large wound, which contact with the air irritated, the pain became violent, a fever was kindled, the large viscera became affected, life was endangered, and often death supervened, mocking the efforts of medical art.

In proportion as the circular limits of the living and

the dead parts enlarged, the suppuration, already fetid, became more abundant, and the softened parts became detached in shreds, bringing with them pieces of the skeleton. The bones remaining in place, deprived of their nourishing ligaments, grew black, and finally fell off of their own accord. This labor of separation by nature, though slow and patient, must be scrupulously respected. If, to hasten it, we seek to shake a single joint of the finger, scarcely held by its eroded ligaments, the wound will not fail to become covered with fleshy granulations of a bad kind, flabby, bloated, bleeding at the least touch, and always threatened with hospital gangrene. Amputations made in the Crimea, on account of freezing, were not successful. M. Boudier, chief physician of one of the field hospital divisions, in one of his reports, attributed this failure to the shattered and wasted condition of the sick, almost all of whom were tainted with chronic diarrhœa.* The cases of congelation happening even in the field hospitals, were a manifest proof of this debility. The inefficiency of the operations imposed the duty upon the surgeon to abstain from them, and to limit his treatment to cleanliness, disinfecting the wound by dusting on it with chloride of lime, and leaving to nature alone, the effort of separating the dead parts. This forbearance of the surgeon proves the fact, that nature draws the line of demarcation between the living and the dead part much better than the hand of the operator, and with much less sacrifice. Art assigns to amputations certain places, which often oblige the sacrifice of portions of a member that might be saved, while nature, entirely conservative, never takes off any part that can possibly live. In observing the operations of nature, we may be convinced that the indication of places called *preferable*, is often theoretical, arbitrary, and not sustained by experience, and that we may amputate precisely at the line of dermarcation between the healthy and the diseased parts.†

* To this cause we may also ascribe the unfavorable results obtained by double amputations, that is, of two limbs at a time.

† The U. S. Sanitary Commission has issued a tract, prepared by a committee of associate medical members, on the subject of amputations

Upon these principles I have now for several years performed a series of partial amputations of the foot. The war of the East has furnished a host of examples which I may use as arguments. Even before I visited the Crimea, I found in the hospital at Marseilles among the wounded sent to France, three hundred soldiers affected with partial freezing of the feet, which were healed, or in process of healing, although art had not intervened, and nature alone incurred the expense of a cure. Nature takes no notice of the decisions of savans in fixing *places of preference.* If a part of a toe can be saved, even when all the others are dead, she saves it. I have seen upon two patients, the second phalanx of the little toe left, after the spontaneous separation of all the other toes; and in others, only the thumb and little finger remained. Nothing is more diversified or more ingenious than the conservative operation of nature. The surgeon must imitate it only, or let her take her

through the foot and at the ankle joint, in which they arrive at the following general conclusions:

"I. In all amputations of the Lower Extremity the surgeon should be governed in the selection of the point of operation, and the method to be adopted,
    1. By the mortality of the operations in question.
    2. By the adaptability of the stump to the most serviceable artificial limbs.

II. In all injuries of the foot, involving parts anterior to the mediotarsal articulation, the surgeon should sacrifice as little as possible of the structures essential to progression. He should preserve
    1. Single phalanges, the importance of which increases from the small to the great toe;
    2. The metatarsus, by amputation of the phalanges, or by the excision of individual metacarpal bones;
    3. The tarsus, by amputation at the tarso-metatarsal articulation. (Hey's or Lisfranc's method.)

III. Of the amputations through the tarsus, or at the ankle joint, preference should be given to Syme's operation as affording a minimum mortality, with a stump best adapted to an artificial limb.

IV. In the after treatment of the amputations and resections above considered, it is good practice to leave the wounds open to heal by granulation."

The above tract has been furnished to all surgeons in the present army service, and is designated as G. in the series of monographs upon surgical and medical specialities, published by the Commission.

course. Let us see how she proceeds: The portion of bone to be separated dries up, becomes black, and projects. At its base, the living flesh swells out, covered with granulations, and encroaches upon the bone, which finally falls off of itself, leaving a deep hole. The granulations fill this cavity rapidly, and thus the stump becomes covered with soft parts, in the best possible manner. The surgeon should not interfere, unless nature is powerless, and even then he should assist with caution, for he has reason to apprehend hospital gangrene. This was the opinion of M. Thomas, physician-in-chief at Constantinople, who observed that the slightest effort to extract a bone, scarcely held by the ligaments, almost always induced hospital gangrene, and he therefore left to nature entirely, the task of finishing the amputation of parts mortified by freezing. It was not until 1856, that the hygienic condition of the hospitals became such, that Messrs. Thomas and Lustreman could practise amputations with success. In the field, the best treatment is fraught with great danger. Latterly Professor Chassaignac has amputated the dead parts of a gangrenous limb, thus relieving the patient of the burden of the lost member, stopped the source of wasting putrid fluids which infected the whole system, and even allayed pain, in relieving by division of the bones, the choking up of the marrow which they contained. This practice is good in common cases, but in the East, it would invariably invite an attack of hospital gangrene.

We see what cares devolve upon the military surgeon in times of war. The soldier has only to evince heroism on the battle-field, and a still greater heroism in obscure labors, and fatigues, and privations, bravely endured; but the surgeon has not only in his own person to share a large part of these pains and perils, but also to seek to relieve those of others; he must also in the midst of active labors, often too severe, work in his mind, to devise the best method for performing one or another operation, and of conducting the treatment, which is ever varying according to the circumstances. Every one knows that the first part of this task was performed by the physicians of the army of the East with devotion, and the army itself proclaimed the sincerity of its regrets, when it saw slowly passing to the cemetery, borne by maimed soldiers, and followed by generals, military intendants, and officers of every grade, the coffin of M. Mercier, chief physician of the field-hospital of the right, who was decorated after the taking of the Malakof, and died two months after in the midst of his sick. Day and night he had remained in the field hospital at Carénage, depriving himself of necessary sleep, until he laid himself down to die.* I wish to demonstrate in a few words that in the fulfilment of the second part of their task, that of scientific observations, we found in our surgeons the same attention and activity.

CHAPTER III.

THE PHYSICIANS.—CHLOROFORM.—EUPATORIA.

HAVING carried my investigations into all the regiments, and taken account of their several organizations, and the causes of sickness and deaths, I often met the surgeons in scientific councils, where we mutually enlightened each other on different subjects, and each, according to his ability, gave his experiences, and profited by those of others. These conferences always ended by a

---

* The funeral honors, due to medical officers in the French service, are as follows:

Inspectors are followed by three detachments, when the death occurs in active service, and by two, when in retirement.

Principals, by two detachments when the death occurs in active service, and by one when in retirement.

Majors, by one detachment, whatever may have been their position at the time of death.

Aides-Major, by a half detachment, whatever their position; physicians and apothecaries commissioned by the minister of war, receive the same honors as the aides-major of regular rank.

In the common salutations of the camp, sentinels "present arms," upon the passing of inspectors, and "carry arms," when a principal or aide-major passes.

session in the amphitheatre. M. Scrive, chief-physician of the army of the Crimea, to which he rendered great services, usually attended us thither. We caused all kinds of operations to be repeated before us upon the dead body by the army surgeons, with the view of ascertaining the most capable, in order to place them as circumstances required at the head of important trusts. A great number of health officers attended daily at these practical reunions. I was often invited to give them advice and examples, which I cheerfully did, and thus not only acquired great influence, but was able to excite among them a laudable emulation, so that they carried even to the battle-field, an ardent relish for study. These sessions proved quite an attraction, and the Sardinian and English physicians, and Sir John Hall among others, honored them sometimes with their attendance. French science was there worthily represented. Under the empire, at least, half of our army physicians had never received their medical degree, and had neither the ability nor the legal right to make prescriptions, and hold the rank of practising physicians. Now, the officers of health are recruited only from doctors of the regular faculty, who are not admitted without being subjected to a new examination. All are educated men, authorized by law and their diplomas to practise medicine.

This was a necessary reform. Does not the military surgeon require a large amount of knowledge and experience so that he shall never fall short of his mission, and be able at any moment to practise the most numerous and grave operations? And yet it is not in the operations alone, however important they may be, that he finds the greatest difficulties to overcome. In amputations, the rules for tying arteries are known and established beforehand; he has had hundreds of opportunities to practise in the amphitheatre, and acquire a certain sleight-of-hand, in such operations. Upon the battle-field the variety and number of the missiles produce at each moment an unforeseen combination of wounds most fearful to manage; here, in place of previous rules, everything must be extemporized. We must act quickly, to save the life which is gushing from the wound. On this bloody theatre, it is not enough to be learned; a quick eye, and prompt intelligence always alive to the occasion, are required. It is this instinctive genius, always so precious and necessary, that I would wish to see developed at the school of Val-de-Grâce, among the students who are to supply the health service of the army. They should receive more instruction in the difficult problems of military surgery, oftentimes reduced, as they are, to the resource of expedients; show them how, with the blade of a sabre, the ramrod of a gun, a bayonet, and even some pieces of a cloak, the apparatus for a fracture may be had on the field of combat. At least they should take a thorough course upon gunshot wounds, and other wounds of war; but, singular to remark, a special chair upon the wounds of war has not yet been established at Val-de-Grâce. It is true this department is not neglected, for the professors seize with eagerness all occasions to initiate their pupils into the practice of military medicine, and the treatment of these wounds; but we may be allowed to hope that the treatment of gunshot wounds, instead of being left to the chance teaching of many chairs, and different professors, will be deemed of sufficient importance to be confided to a special teacher. Already Marshal Vaillant, the minister, with whom the soldier's health is an object of constant care, has endowed Val-de-Grâce, since June, 1857, with a special chair of diseases and epidemics of armies. The department to which I allude, would add a new lustre to the reputation of Val-de-Grâce. It is true, that to fully comprehend the importance of these special lessons, and to illustrate their bearings upon the higher departments of science, a long course of practice is necessary on the field of actual warfare; but the pupils of Val-de-Grâce would find at least in the writings of our illustrious predecessors, especially Percy and Larrey, a host of practical illustrations, a rich nomenclature of the incidents of war, and ingenious methods of treatment; and they will learn how a surgeon may overcome every difficulty, which war continually offers.

These difficulties are innumerable. Let us remember,

however, that in the Crimean war, surgical science was first aided by a recent discovery due to the researches of M. Flourens, and which until then had not been used on the battle-field. We allude to the anæsthetic action of chloroform, whose wonderful effect in relieving the terrible pains of the wounded, has been often useful in healing their wounds. Its use permits us to trim wounds, mortal in appearance, which the surgeon would not have ventured to treat with so much energy, for fear of exciting new sufferings. These wounds being thus treated, become less painful, and sometimes we are astonished at their unexpected cure. For example; a soldier of the 57th regiment received in the upper part of his thigh a piece of a bomb weighing 2 kilogrammes, 150 grammes (4.74 lbs.). This enormous piece of iron buried itself so deeply that only a small projecting corner could be seen. Chloroform permitted the extraction of this mass, and afterwards amputation, without the patient suffering the least pain, and he rallied. Without this agent, we should have hesitated to attempt the operation.

By subduing pain, chloroform gives a calmness and mental tranquillity to the wounded, very favorable to healing. It deprives the traumatic fever of an excess of reaction, often caused by the anxiety of the patient. Before the discovery of this precious agent, some soldiers, it is true, endured amputations without a groan. I have seen an Arab continue to smoke his pipe, while I took off his arm; but this paroxysm of courage, gained by a great exaltation, fell some days after into a nervous depression, so much the more dangerous on account of the stoicism which had hitherto sustained him. Therefore, those surgeons who know how dangerous it is to struggle against suffering, tell their patients not to resist it, but to let it express itself in cries. I remember an unfortunate soldier, who, having decided to allow his leg to be cut off, exclaimed: "They told me it was nothing—that some sing under the operation; I should like to see them!"

We know that, in great operations, death ensues oftener from nervous prostration, consequent upon excess of suffering, than from hæmorrhage. It is the same with animals. M. Claude Bernard observed, that in laying open the spinal column, in experiments upon rabbits, they always died at once, unless they had previously been rendered insensible by chloroform.

But used imprudently, this agent, while it takes away suffering, may also take away life. It shares this sad privilege with the most potent remedies; taken in large doses, they are for the most part poisonous, and kill instead of curing. The danger may perhaps be certainly avoided, by observing certain rules, and especially by not pressing the inhalation to its extreme limits. This extreme limit is, in my opinion, when in accordance with the precept laid down for several years, we pass the stage of *insensibility* and reach that of *collapse*, and a complete muscular prostration. This condition is reached, when a limb is lifted up, and falls like a mass of dead matter, for then, life almost borders upon death, it has retired into the vital centre, placed by M. Flourens in the medulla oblongata, at the origin of the eighth pair of nerves which absolutely control the function of the heart and lungs. To approach this point, is rash temerity: to reach it, is death. This rashness appears to me unjustifiable. We ought always to stop when insensibility is reached; it is enough for the patient that the pain is alleviated. The surgeon may find it convenient to cause a prostration of the vital powers, but should abstain from affecting it by such dangerous means. The movements of the patient are easily restrained by the assistants, and if not entirely repressed, I would even then abstain from its use.* The physicians of the army of the East were of this opinion, and administered chloroform with great prudence, stopping at the point of insensibility, and never intentionally exceeding it. They had therefore no fatal accidents to deplore, although in the campaign of the East this agent was employed at least thirty thousand times. In the Crimea alone, it was administered

* I have given this precept all the proofs necessary in a memoir read at the Institute, July 19, 1853.

to more than twenty thousand wounded, according to the estimate of M. Scrive. The physicians of the Sardinian army, at the beginning of the campaign, hesitated to use it, but the success of our surgeons soon gave them confidence in its efficacy. Henceforward we may have a steadfast confidence in chloroform, and thank Providence for having allowed human skill to invent an agent that can suspend pain.*

Epidemics made such ravages among our physicians, that after the taking of Sebastopol, the field hospital of the left division was converted, upon the requisition of M. Scrive, into a special home for convalescents belonging to the personal service of this department. It was located upon the heights of Sebastopol. It had its share of sorrow. A little lamp lighted for use at night, was observed by the Russians, and made a target for their guns. Their bombs crashed through the roof. We hastened to carry away the sick upon mule-litters, and when the danger had passed they were brought back.

In front of this hospital a magnificent panorama spreads itself. Towards the south, bounded by a horizon of vast extent, the sea was covered with the supply ships of the allied armies; to the north a splendid road without rocks or sandbars, finely sheltered from the winds, and easy of access, was dotted with vessels destroyed by the hand of man; and at the head of the harbor, the granite docks and marine railways, masterpieces of human skill, which were going to be

---

* The English surgeons, in the Crimea, employed chloroform very generally, and McLeod considered their confidence in its efficacy greater than among the French. Only one well established fatal case occurred from its use, while a great number of operations were successfully performed which would have otherwise not been attempted. The *morale* of the wounded was better sustained, and the courage and comfort of the surgeons increased. Its moderate use was not found to result in depression, but on the contrary it often supported the strength of the patient under the operation, and it was never more successful than when used immediately after the injury, before the constitution had begun to suffer from the nervous irritation liable to follow a wound. It acts more rapidly upon persons who have lost much blood, and hence more care is necessary in such cases.

destroyed, challenge our wonder, and suggest fruitful themes of reflection. Beyond these, arose three formidable lines of Russian batteries, while still beyond lay the camps of the enemy, spread over a plain of great extent; to the east irregular mountains, commanded by the bastion of Malakof.

As for Sebastopol, this city, lately so proud and so menacing, presented only a spectacle of mingled ruins and tombs, with great heaps of disabled cannon, and broken gun carriages, balls, shells, and bombs. In walking about it was necessary to avoid the channels swept in enfilade by the Russian artillery, and to thread our way with difficulty, through the encumbered and obstructed streets cut in the steep ridges of this promontory, which could not be attacked in front. The house occupied by General Levaillant, Governor of Sebastopol, was not bomb-proof, as was made evident by large openings in the roof. The general led me to a corner of his little garden, where he had built his observatory, and lent me his spy-glass. We could observe very distinctly, the Russians on the other side of the harbor, loading and pointing their cannon at us; but their shot passed over our heads.

These shot often committed ravages, through the carelessness of our soldiers, who exposed themselves rashly and needlessly; they had seen so many of them, that they took no care to avoid them. This negligence added new wounded to our field hospitals; but epidemics were the chief agencies in filling them. Surgical operations diminished and gave place to therapeutic treatment. The sick, coming in by crowds, encumbered our hands, and obliged us to send to the regular hospitals all cases of an obstinate character. A field hospital admits readily of expansion, according to its wants; it is only necessary to bring into use more tents, houses, or barns, as they may be needed from day to day. It is necessary, however, to hasten as much as practicable the distribution of the sick among the various hospitals, in numbers of five, or at most, six hundred men to each. If this number was exceeded to double or treble, as it often happened, we ran great risk of miasmatic infection.

This peril is encountered even in times of peace, when we accumulate at a given point a large number of sick; and in extensive hospital establishments, the relative mortality is much greater than in those of small extent, notwithstanding every hygienic precaution. This truth is not generally admitted, except by medical men, and it is usually held that, by a system of centralization, the service is better attained; we violate the laws of health, prolong the stay of the sick in the hospitals, increase the number of invalided and of deaths, and generate infectious diseases. With the view of emptying the field hospitals of the Crimea, we opened in December, 1855, some new hospitals at Constantinople. I went to visit them, after having continued, by way of Eupatoria, my tour of inspection.

Eupatoria is a large city, which the Russians would have burned had they found time, but we saved it from ruin. Its houses are mostly one story high, and very spacious, especially those bordering upon the larger streets. The principal thoroughfares are much cut up, and very muddy in winter, but they are bordered by sidewalks about twenty inches above them, which are sheltered by overhanging roofs. At this time the city was occupied by the Allies. Only a poor part of the population remained. The wealthy had retired, but the poor were increased by the daily arrival of the unfortunate villagers from the country around, who said they had been burnt out by the Russians. They were not only received, but fed by the Allies; I have seen biscuit distributed to more than a thousand children, who in return performed such services as they were able, for the corps of engineers. The sanitary condition of the French troops was very satisfactory, and we counted but 300 sick in an effective force of 12,000 men. Such a result could not be attained, even in France; it is due to the continuance of fine weather, an abundance of provisions of good quality, which were regularly distributed; and finally, to the military evolutions that were executed, which preserved the soldiers in a high state of morale. The corps of Generals D'Allonville and Failly, which harassed the enemy, showed very few sick upon their rolls; and the pleasure of burning ammunition, combined with their successes, soon transformed the young recruits into veteran soldiers.

These corps were encamped under shelter-tents, which would become unserviceable as soon as the first rains should soak the ground. Would it not have been better to lodge our troops in the city itself? It was showing too much respect for the property of our enemies; the inhabitants would have been asked neither beds nor blankets, but only a roof to shelter the men against the rains, and the rigors of a winter which we knew would be inclement.* There was an abundance of houses. The diseases, it is true, were not serious, but the dangers of an unhealthy camp might increase their number, and enhance their severity. In anticipation of this, the field hospitals were arranged under sheds in an immense area, inclosed with walls, and having for their dependencies three fine houses which might afford accommodations for two hundred beds. It would be easy to prepare tents in the courts, if needed. The physicians were assembled in conference, various measures were discussed, and attention was called to the Turkish field hospitals, from whence the germs of epidemics might perhaps arise. The 15,000 Turks and Egyptians, assembled at Eupatoria, had many thousand sick, chiefly of scurvy, which they sent to Varna. Their field hospitals were well arranged, but unfortunately were attended by few physicians worthy of the name. It was agreed with Achmet-Pacha, General-in-Chief, that certain effective measures should be taken, under the supervision of M. Bourgillon, chief physician of our hospital, and one of the most distinguished of our medical officers, who would thus learn the peculiarities and sanitary condition of the sick in the Turkish and Egyptian armies. M. Bourguillon was highly gratified in the relations with M. Cassini, chief physician of the Egyptian army, who conducted a difficult service with much ability.

* This suggestion, made in my report to the Minister of War, and to Marshal Pelissier, was attended to, and the army took up its winter quarters in a portion of Eupatoria.

Our Mahomedan allies, less scrupulous than ourselves, seized for their use all the houses of consequence, even the mosques, and the magnificent Hebrew synagogue, which was one of the finest in the world. They, however, gave up to us, of their own accord, two large houses, well furnished with hospital furniture.

In the Turkish field hospitals, as in the English, the physicians enjoyed great authority in their administration and control. Their attendants were soldiers of long service, but too young to be pensioned for retirement, who generally evinced much zeal, because the loss of their place would deprive them of all right to a pension. The food was composed largely of mutton and chicory. In the East, mutton is abundant and of good quality, but beef is scarce, lean, and poor. Encouraged by this example, I desired that mutton and chicory should sometimes be distributed to the hospitals instead of broth and boiled beef, deeming it proper that in distant countries we should avail ourselves of the resources of the locality, and adopt wise modifications in our regimental customs. We borrowed another practice from the Turks:—that of frequently fumigating the wards of the sick, by throwing dried sage upon a pan of burning coals. The aroma thus occasioned, is agreeable; it rapidly and completely renovates the air contaminated with miasms, and if we open for a moment the doors and windows, the fumes soon escape. This ancient mode of purification is not to be despised.*

The want of Medical Science, which the Turks have long felt, will be soon supplied, as the Sultan has founded

---

* The English Sanitary Commissions were not favorably impressed with the results of fumigations. The deodorizing substances used, diminished the odor, but not in the same ratio the disease; and they expressed decidedly the opinion that they should never be trusted to for protecting health, if it be possible to remove the nuisances at once and to a distance. They observe that "burial of putrid refuse to a sufficient depth, when removal is impossible, or when the substance cannot be destroyed by fire, is a safer expedient than removing its smell by charcoal, or by any other deodorizing agent scattered over its surface. Smell is indeed the natural index to danger, and removal or destruction of the offensive matter is the remedy. There is reason to fear that after the smell is removed the danger remains."—TR.

at Constantinople, a school of medicine, in which five hundred pupils are collected. The more intelligent are to be sent to Paris to finish their studies, and this nursery for the education of young men will confer great services upon the Ottoman army, and spread among their co-religionists our ideas and our customs.

Upon leaving the Crimea, I went to inspect not field but general hospitals. The first condition of the latter is their permanence. At Constantinople, these vast establishments increased and multiplied daily. They were far from the enemy, and were sheltered from all danger from abroad. They were near enough to the Crimea for easy communication; they were on the seashore, and the means of transportation were easy. Their organization was excellent.

Shut out, by the blockade of Sebastopol, from transportation by sea, the Russians were compelled to remove their sick in four-wheeled waggons, without springs, drawn by three horses, and arranged for four patients, of whom the two that were best able to endure it were seated upon a kind of bench, while the other two lay upon straw. Rude as this conveyance was, it proved very acceptable to the Russian soldiers, and by this means of conveyance they were sent to Baghtchehsaraï and Sympheropol, where 15,000 beds were arranged in various hospital establishments. They underwent great exposures, and often suffered intense cold upon this long journey, and the mortality among these unfortunate men was very great.

The English waggons, used for moving the sick, were suspended upon good springs. Two beds laid upon litters were placed in the carriage, after the manner of drawers, and the sides were made of light rails, to insure the circulation of fresh air. Such men as could travel without lying down, sat in a kind of covered seat placed in front, and others even rode upon the top, when the weather permitted.

Our ambulance waggons were well hung on springs, and proved a great luxury. A covered seat in front received such of the wounded as could sit up, and in the other part, arranged like an omnibus, two sick could lie upon litters placed on the sides. The seats were elastic, well cushioned, and covered with red sheepskin leather. These carriages were heavy, and but little used, as we preferred to use mules, provided with chairs or litters. In a campaign, the simplest measures are always the best. Nevertheless, although these carriages were very comfortable, it was of great advantage that their use was confined to removal to short distances; to the infirmaries and field hospitals, and from thence to the place of embarkation. The fleet, whose business it was to supply our army with necessaries of every kind, rendered also great assistance in transporting from the Crimea to Constantinople, many of the sick who could never have recovered their health in the Crimea.

PART III.

THE HOSPITALS AND THEIR DISEASES.—
TYPHUS IN THE CRIMEA.

It was not against the Russian arms alone that the allied troops of the Crimea were forced to struggle. Those acquainted with the history of long campaigns are aware that accidental or endemic diseases commit greater ravages among soldiers, than gunpowder and the sword. Besides the hygienic precautions necessary to preserve those in health, and the care demanded by the wounded, the wants of the sick and convalescent press incessantly upon the military administration the most painful problems of medical science. If we review the history of our hospital establishments during the war in the East, this fact will be shown; and I trust that the administration and science will never cease from their exertions until these problems, which are the highest aim of their double task, are solved.

CHAPTER I.

THE CHOLERA.

It will be remembered that, at the beginning of the war, Gallipolis was selected as a place for assembling the various contingents arriving from the several ports in the south of France, and from Algeria. This peninsula was to be made the strategic point of the army of the East, and the base of its operations. By the active forethought of General Canrobert, it has been rapidly converted in reality into a military station, devoted to camps and provision stores of all kinds, including materials for regular and field hospitals. Each had upon the front of its color line, its particular guidon, and each had its separate cantonment. As new regiments debarked, they pitched their tents upon undulating but elevated grounds, whose salubrity, recognised in advance, is preserved by the fresh sea breeze. The energies of the medical corps of the army were directed to putting into force certain sanitary measures, which were also applied to the city of Gallipolis itself. They had to struggle against the well known indolence of the Mahommedans, before they could obtain the removal of heaps of rubbish and filth, that endangered the health of the town. In oriental cities, this duty is left to be discharged by the sun and winds, the former having for its share to calcine the heaps of filth, and reduce them to powder, when the winds are expected to carry them away. The horrible stench of these accumulations of rubbish seems a standing invitation to epidemics.

While the brigades were being organized, the old athletic and bronzed soldiers, who had served in Algeria, sought to initiate their comrades into martial usages, and as war was something new to them, to accustom them to the habits and life of the camp. They undertook to teach them, according to their picturesque expression, *how to handle the tools*, that is to say, how to endure hardships like themselves, and how to be prepared to practise the art of resisting the inevitable privations of the campaign, and at the same time to retain their health. General Canrobert did not leave the troops inactive, but accustomed them to the fatigues of war, by making them labor upon earth works, and construct immense trenches around their camps, rendering them as capable of defence as forts. With the assistance of the English army, they fortified the peninsula of Gallipolis by entrenchments which extended from the gulf of Saros to the sea of Marmora; these works were to close against the Russians the passage of the Dardanelles, which they had opened in 1829. Besides being useful in a military point of view, they served a sanitary purpose, with the most happy results. The number of the sick at Gallipolis was not large, and most of these had only slight indispositions, which kept them but a few days in the field hospitals. A hospital with three hundred beds, established a kilometre (0·62 miles) from the city, took the place of some houses in the town which had been temporarily occupied by our sick, and abundantly sufficed for the first necessities of the army. This first hospital barracks of the French army was established in May, 1854. Placed near the route of the fleets, upon the shores of the Dardanelles, at a point where lighters could easily land, it was used by us, after the departure of the armies, with the greatest advantage. At this hospital, the sick on their way to France from the Crimea and Constantinople were left, when it was found they could not encounter the voyage without danger. It therefore became an appendage to the hospitals at Constantinople.

We had at first committed the error of constructing the barracks upon low grounds, with the view of utilizing some ruins, and to be near a spring; but this fault was subsequently remedied, when it became necessary to extend our hospital resources. At a place but fifty yards distant, we found an elevated and well aired plateau, where we prepared several barracks, with beds sufficient for 300 sick. The hospitals at Gallipolis, when thus completed, had 600 beds, and were always well managed by the intelligent and devoted medical attendants under the care of Dr. Molard. I found the litters and furniture in complete condition, and the bread, wine, meat, and broth of good quality.

But the events of a campaign advance with rapidity; and when once assembled, our troops did not remain long at Gallipolis. Nearly a hundred thousand Russians, followed by large reinforcements, laid siege to Silistria, which was bravely defended by eighteen thousand Turks. The troops of Omer-Pacha amounted to a hundred thousand combatants, but they were scattered at many points, chiefly at Routschouk, Silistria, and Chumla. This barrier might be overthrown at any moment by the invading army, and it appeared urgent that we should hasten to the assistance of the

Turks, and place Adrianople in condition for defence by a prompt movement. Each movement of the army required the creation of new centres for hospital service.

On the 7th of May, 1854, the Marshal Saint Arnaud arrived at Gallipolis, reviewed our enthusiastic army, and left his instructions to embark at once for Constantinople, to which place he went himself the next day. He imparted his own activity to all whom he approached, and his lively and animated tones excited life even in the Ottoman authorities. The sultan himself confided in the marshal, and ordered all the resources of his empire to be placed at the disposal of the generals of the allied armies. Activity succeeded the slowness and hesitation of the Ottoman administration, so accustomed to leave everything till the morrow. On the 19th of May the marshal and Lord Raglan went to Varna, held a conference with Omer Pacha, passed in review at Chumla a corps of 45,000 soldiers of tried valor, and decided to send forward, not a single division as he had at first proposed, but the whole available forces at his disposal. Varna then became a new base of operations, which left Gallipolis a secondary point. Large supplies of provisions, ammunition, materials of war, and hospital supplies were hastily sent forward. On the 1st of June, 6000 troops, composing the first brigade of the division of Canrobert, embarked from Gallipolis, and an English detachment of equal force at the same time from Scutari, where Lord Raglan had his head-quarters, and advanced by land to Varna, about twenty-three and a half miles distant. Other French regiments were to arrive by sea and land, at the common rendezvous.

On the 11th of May, a barrack commission, in which Dr. Cazalas represented the medical interests, assembled at Adrianople, the ancient residence of the Turkish sultans, and the capital of Roumelia. This city, by the beauty of its climate, its wealth, its resources of every kind, and especially its location, which commanded the Balkans, and the route which the enemy must necessarily take, became a strategic point of the greatest importance. They hastened to place at our disposal an immense barrack, built in 1820 by order of the Sultan Mahmoud. It formed a parallelogram about 490 yards long from north to south, and 300 yards wide from east to west, built with a ground floor and one story. The angles were relieved by square four-story towers, each surmounted by a gallery and terrace, with a staff from which floated the national colors. In the centre of the principal arcade was the entrance to the pavilion of the sultan, quite in oriental style. It was supported by many storied ranges of white marble columns, among which the winds played freely, and had a grand portico of sculptured marble adorned with gilded arabesques. Five large marble basins, each furnished with twenty large copper spouts, and fed by an aqueduct, furnished an abundant supply of good water. There was a wide difference between this monument of taste and our barracks in France, where rigid economy allows no freedom to the inspirations of the architect. The premises we have described could lodge 10,000 soldiers, and contained 278 rooms receiving light from 1280 external windows. It was at once decided that a hospital for 1200 sick should be established in a part of this magnificent barrack, and to remedy as much as possible the inconvenience of so great a concentration of sick it became necessary to assure each patient at least 45 yards of air. In our hospitals the common measure is from 24 to 26 yards, and in our barracks from 16 to 18 yards. Subsequent events having rendered Adrianople less important as a military point, we reduced the number of beds to 300.

On the 16th of June, when the division under General Bosquet, amounting to 11,435 men, and the troops of General Morris, composed of about 1200 cavalry, arrived at Adrianople, a hospital was prepared to receive 169 sick and 250 lame. The former division left for Varna on the 25th of June, but the cavalry regiments of General Morris did not follow, and at a later day they left their bivouacs on the plain of Tundja for the barracks where they passed the winter of 1855.

The first French hospital established at Constantinople was that of Maltépé, and the first sick that were received belonged to the third division, commanded by Prince Napoleon. This division, having left Gallipolis May 28th, proceeded by land along the borders of the Sea of Marmora to Constantinople, leaving their sick and lame about half way at Rodosto, in a transient hospital arranged for 250 beds. They would have retained this as well as the barrack accommodations used by the victorious Russians in 1829, had not the siege of Sebastopol been decided upon. On the 7th of June, the third division entered Constantinople, and bivouacked on the plain of Daoud Pacha, leaving upon the minds of the Turks a vivid impression of admiration and wonder. They saw with surprise the oriental costume of our Zouaves, which had been abolished among them by a reform, against which the old Ottoman party had protested; they still wore, by a kind of tolerated disobedience, the ancient national costume.

Maltépé was a Turkish hospital, one half of which was opened June 7th for a field hospital of the third division, and the remainder some months after. At a distance of some 2000 yards from the chateau of Seven Towers and the strong walls of Stamboul, on the west side, appear in strong profile upon the poetic sky of the East the two great permanent barracks of Daoud Pacha and Ramis-Tchiflik. They were modelled after those of Adrianople, and were distinguished by an architecture in which elegance did not detract from solidity. They were about a mile and a quarter apart, upon an extensive plain deprived of trees, but covered in summer with rich harvests. The hospital of Maltépé was built between the two barracks upon a little eminence exposed constantly to the sea breeze, and of sufficient capacity for 450 sick.

The third division was reviewed on the borders of the rich vale of the tombs of Eyoub, by the Sultan and his brilliant staff; and on the morrow (June 18th) it embarked for Varna. The field-hospitals followed this movement, leaving their sick at Maltépé, where the suffering soldiers sent from Rodosto were received.

From this day, a hospital was definitely arranged, at the head of which was the learned chief physician M. Durand, who continued in charge during the campaign.

The sick arriving by sea, were landed at the foot of the Golden Horn, from whence the convalescents walked, while others were transported upon litters, chairs carried by mules, or ambulance waggons. The road was very narrow and steep as far as the Porte-des-Canons; the feeble were often scarcely able to make the journey, and the Turks, a people considered in Europe as wanting in compassion, assisted them or made them sit down. Upon reaching the external walls at the Opening of the Crosses, the route continued to rise by a very gentle slope to Maltépé. It passes through the immense cemetery, planted with resinous trees and the common cypress, which borders the long western line of the ramparts of Stamboul. We soon arrive at a little historical hillock, on which is a small windmill, the only one on the plain. It is here the troops of the Sultan Mahmoud were harangued by their chiefs and ulemas in 1828, when departing for Maslak, where they massacred in their camps the revolted janizaries. The hospital is 200 yards beyond, and from its principal front is displayed a beautiful panoramic view of Constantinople, the Sea of Marmora, the Princes Islands, and the Mountains of the ancient Bithynia, capped with snow. The sick were never tired of admiring this fine spectacle, which inclined them to a calm and quiet meditation, so salutary in convalescence.

The establishment at Maltépé formed a great rectangle, and the four sides inclosed a court of very large extent planted with trees. The walls were of wood on the side towards the court, but of stone externally, and the side towards Constantinople had only a ground-floor surmounted at the angles by little turrets. Externally it was bordered by an orchard, cooled by refreshing springs received in marble basins, and had in the centre a monumental entrance-way of white marble in good Byzantine style. This range of buildings contains several of the hospital accessories, such as the Turkish baths, the laundry, the kitchens, the apothecary rooms, the offices and two chambers of honor, one called the Sultan's, and the other the Seraskier's, or Minister of War. The other three sides of the rectangle contained a ground-floor, and one story above it, along which, on the side towards the inner court, ran a corridor to give access to the rooms which looked on the country. Each chamber contained from 30 to 40 Turkish beds, which were made of large fir boxes sustained by iron legs, and containing two mattresses of cotton or wool. An aqueduct, abundantly supplied, distributed excellent water through the whole establishment. The artificers of the engineers immediately arranged everything to our wants, so different from those of the Turks, and this hospital continued to be used by us constantly until May 31, 1856, when the Crimean troops began their embarkation for France, which was completed on the 5th of July following, under the personal observation of Marshal Pelissier.

Meanwhile the ranks of the army, composed of some 15,000 to 20,000 men, increased from day to day, a fourth Division having joined the expedition, and soon after a fifth. They all proceeded in succession to Varna, whose ramparts are situated at the foot of a marshy valley, between two spurs of the Balkan range. It was taken by the Russians, in 1828. The city contains 16,000 inhabitants, and has on one side an extensive lake, and on the other the sea. Its harbor is very difficult of access, and offers but little shelter and very poor anchorage. Upon their arrival the regiments were stationed about five miles beyond, where they pitched their tents upon a high plateau called Franka, approached by paths winding through most beautiful gardens. The elevation of this plain was about 650 feet above the sea, its breadth from east to west was about two miles, it was bordered by forests, and refreshed by many springs of limpid water. The Russians had bivouacked here in 1828; it was in all respects most suitable for the encampment of an army. From this point our troops watched the defiles of the Balkan mountains, escaping the noxious miasms to which a sojourn in the low grounds would have exposed them. Unfortunately they committed indiscretions; eat unripe fruit, and drank immoderate quantities of spirits or cold water. In the evening, when the heat of the day had fallen from 30° to 12°, and even 10° Centigrade (86°, 54° and 50° Fahrenheit), they would often remain in the open air, and sleep without their clothing, being in this respect more careless than the inhabitants of the country, who avoid going out of doors after nightfall, unless protected against the low temperature and humidity by warm and impervious clothing.

Although the general sanitary condition was satisfactory, yet a number of men entered the infirmaries, sick with intermittent fevers, and especially with the intestinal fluxes, precursors of cholera. It became necessary to create asylums for the suffering troops. The Ottoman authorities placed at our disposal a very large permanent barrack, which we shared with the English. It received 700 beds complete. The buildings were old and in bad condition. We contented ourselves with making the most necessary repairs. It continued during the campaign to receive soldiers from the Crimea, and chiefly from Eupatoria. Besides this permanent hospital, there were established upon the elevated grounds several large field-hospitals, two of which were exclusively reserved to the Cholera patients from the fatal expedition of the Dobrutcha.*

* The English suffered from cholera on two occasions during the Crimean war. The first epidemic appeared in April, 1854, increasing for three months, and again declining slowly, and in an irregular manner till February, 1855, when it disappeared. The following month marks a clear interval between the two epidemics, for no case occurred in the English army in March, 1855.

The second outbreak began in April, 1855, culminated in June, and declined slowly and irregularly as in the former. It was less severe than the first, but showed a tendency to remain permanently, and after a month's interval again increased, as if to make a third annual visit. The total number of cases that occurred, was 7,575, and of deaths 4,513 or 59·57 per cent. of the deaths. 2,902 were in the first, and 1,611 in the second period. The cavalry lost 399, the ordnance 460, and the guards and infantry 3,654. Among the prophylactic measures adopted, were the discontinuance of parades and avoidable fatigues, a frequent change of encampment, strict attention to cleanliness in the camps; suppression of unripe fruits, and unwholesome wine and

The whole coast from Varna to the Danube is a desolate country, covered with steppes and marshes, whose neighborhood is fatal during the heats of summer. In the spring of 1854, Omer Pasha said to Commandant Henry, envoy to his camp at Chumla, " If the Russians remain another month in the Dobrutcha, their army will be annihilated:—to me it would be equivalent to winning a great battle." The terrible havoc made by malignant epidemics in the Russian army in 1828, could not be entirely forgotten; and without doubt, the remembrance of this, in part decided the Russian generals to quit the Dobrutcha and to reascend the Danube to Silistria; and which made them abruptly quit this place after a futile but not fruitless siege. The city, breached on every side, was on the point of yielding, and the vigor of its defenders was on the point of giving in to the numbers and desperate valor of the assailants. The retreat of the Russians on the left bank of the Danube gave to the allied armies, who were impatient to march to the combat, a mingled feeling of painful surprise and almost of discouragement. Marshal Saint Arnaud felt it was necessary to produce a powerful reaction in the morale of the army, employ his soldiers, draw them from fatal inaction, and revive their ardor by some great movement, whose boldness and wisdom should draw the attention of Europe. At that time, the Cabinet of St. James earnestly insisted, that we should proceed to the Crimea, destroy Sebastopol, and annihilate the Russian fleets in the Black Sea. The instruc-

spirits; the furnishing of a spirit-ration, and an increased quantity of fresh meat, and a portion of barley or rice daily on medical representation; coffee for men returning from guard-duty, and every effort to induce men affected with diarrhœa to report themselves for medical assistance in the early stages of the complaint.

It was particularly urged to use remedies in the early premonitory stages; but with their best efforts the influences of climate, vicissitudes of temperature, heavy dews, defective diet, impure water, unripe fruits, and sour wines, operated against the best devised measures for relief, and defied the control of the medical staff.

The greatest mortality occurred at Varna. The period of greatest intensity was in Dec. 1854, when 888 cases in an average strength of 29,727 appeared, and 636 died, giving 2·9 per cent. of deaths to the whole force, and 71·6 per cent. of admissions. —TR.

tions to the Marshal Saint Arnaud were less imperative, and left him free upon this point. The exploration of the shores of the Crimea was therefore commenced, and as soon as the enterprise was considered to be practicable, it was irrevocably decided upon, notwithstanding the opposing advice of the admirals of the fleet, who distrusted the sea at this advanced season of the year.

In the midst of preparations, occasioned by the prospect of early departure, the unmistakable signs of Cholera took our army by surprise. On the 9th of July, the pestilence appeared in the hospitals at Varna, although the relative number of all the sick at the time was not over 600, to the 50,000 men already collected; it was doubtless imported into the East by the successive contingents of the fifth division, embarked in the South of France, where the population was a prey to this epidemic. It first made its appearance at Pirea, and then at Gallipolis, where in a few hours it took off generals Duke of Elchingen and Carbuccia. The expedition of the Dobrutcha was not slow in furnishing new victims. Great as was the wish to transport the allied armies at once to the Crimea, it could not be done before fifteen days, this delay being necessary to make the preparations for departure; it was determined to profit by the opportunity, and make a demonstration that should annoy the enemy, and deceive them as to the plan of attack on Sebastopol.

According to the official reports of a colonel of the staff, sent upon the ground, the Russians had at 45 leagues from Varna, near Babadagh, 10,000 men, with 35 pieces of cannon. The three first divisions of the French army were sent to find them, and they were directed to proceed along the sea-shore, for the convenience of provisioning. It was thought that a frequent change of bivouac would lessen the choleraic tendencies of the troops. On the 21st of July, General Espinasse, who commanded for the time being the first division, while General Canrobert was exploring the shores of the Crimea, received orders to move upon Mangalia, at the head of 10,500 men, of whom 328 were officers. There were left in the infirmaries and hospitals at Varna 16 officers and 925 soldiers. The 1st Regiment of Zouaves, transported by sea to Kustendjé, was to operate at the head of the column under the orders of General Yussuf, and to support 2,000 or 3,000 Oriental cavalry, organized from the undisciplined bands called the *bachibozouks*. The physician-in-chief of this division was M. Cazalas, an energetic man, who had evinced a thorough knowledge of his profession at the school of Val-de-Grâce. He was assisted by several chosen physicians, Doctors Quesnoy, Bailly who was taken off a few days later with the cholera,—and Raoul de Long--champ, who resisted the pestilence as if by a miracle. The means of transportation intended for the sick, consisted of 65 pairs of mule-chairs, 5 pairs of litters, some ambulance caissons, and a number of arabas, loaded on starting with provisions for ten days, for the use of the expedition. Each soldier carried besides provisions for five days.

To accomplish the six and three-fourth miles of the first stage from Franka to Kapakli, the soldiers remained ten hours upon foot, exposed throughout the day to a heat of 30° Centigrade (86° Fahrenheit). In the same evening four cases of cholera appeared. On the 22d, they started at four o'clock in the morning, and the division did not reach Tchatal-Tchesmé until seven in the evening. They travelled about eleven miles, the thermometer rose to 35° C. (95° F.), and the march was very difficult, owing to the route passing over steep and gullied places; but beyond this bivouac, the column descended into a naked plain, without trees, about 124 miles in length, covered with high herbaceous stalks, many of which lay rotting upon the ground, where they formed a tangled mat, through which the troops advanced with great difficulty. This was the Dobrutcha, covered with lakes and marshes which infected the atmosphere, especially in this season of the year. Geographers have placed it as between the Danube, and the walls of the camp of Trajan, about five miles from Kustendjé, but medical topography would place its southern limits as far south as beyond Kavarna, where the troops arrived three days after leaving Varna. The encampments which marked the subsequent stages of the expedition were all equally insalubrious. At Sattelmuch-Gol, Mangalia, Orgloukoï, and even at Kustendjé, as in the ruins of the village of Kergelak, we found no place of encampment other than low marshy grounds, whose waters were poisoned with vegetable substances in process of decomposition. As we approached the Danube, we met some parties of Cossacks, who opposed no serious resistance; the aspect of the country became more and more desolate, cultivation disappeared, and every trace of human industry was lost. We met only some shafts of broken columns, and tumuli of great antiquity, silent witnesses of former civilization against modern barbarism. Since the Russian invasion in 1828, this region, fearfully ravaged, had become almost deserted. Clouds of vultures followed our steps, to devour the dead bodies which might be left behind; it was not unusual to meet troops of wild horses, who took to flight on seeing us. Some shepherds, whose constitutions present the characteristics of paludal cachexia, are almost the only inhabitants of the Dobrutcha; and they are forced, like the beasts that they tend; to use foul water, dipped from the pools, cisterns, or abandoned wells.

The army had furthermore to endure tempests of rain, and numerous atmospheric vicissitudes of heat and cold. The cholera, which until this time had remained almost inoffensive, no longer delayed to make a sudden and fearful attack. On the night of July 30, 300 Zouaves were violently seized, and the *bachi-bozouks* were also terribly assailed by the epidemic. General Yussuf prepared to march forward, but the redoubled strokes of the pestilence compelled him to fall back. His troops scarcely found time to bury the dead that fell by the way. He, however, in spite of all difficulties, transported upon gun-carriages and upon horses, the sick, whose number increased every moment with frightful rapidity. The column of General Espinasse, hoping to surprise the enemy, had made a night-march on Kargualeck, without knapsacks, and with very few waggons, in order to proceed more rapidly, and arrived at about one o'clock in the morning. It started in a rain-storm which lasted several hours. The following morning witnessed a sorrowful scene, for instead of the enemy, they saw 300 soldiers belonging to the 1st regiment of Zouaves, crowded into the mean village ruin, and struck down with the cholera. Means for transportation were wanting. The 9th battalion of foot-chasseurs, forming the rearguard, showed a remarkable example of devotion. They took up all the sick upon litters, prepared on the spot with their shelter-tents, and poles found in the village. The column, ravaged by the pestilence, retraced its steps towards its former bivouac near the great lake of Pallas. It was compelled to leave in a field-hospital, until the morning, a large number of cholera patients, which it could not carry away. On the 31st of July, the surviving portion of the division arrived at Kustendjé, where it found the houses full of *bachi-bozouks*. About 1800 cholera patients were awaiting their turn to embark upon steam frigates, and 1,200 dead bodies were buried in trenches dug around the place.

The unexpected arrival of General Canrobert at Kustendjé, gratified the earnest wish of all, and produced a keen and touching emotion. He called a medical council; impressed new energy into the measures already taken by General Espinasse, who was then suffering from cholera, and revived the courage which the disease had checked. The division, making unheard of efforts to transport the cholera patients which fell from time to time, arrived at Mangalia on the 3d of August, where the forethought of General Canrobert had brought together resources of every kind, and especially fresh provisions, wine, brandy, coffee, and sugar. New deaths were counted by hundreds, and 2,000 sick were embarked for Varna. The marshy stopping-place of Mangalia was rendered still more dangerous by the putrid decay of many dead bodies, which the *bachi-bozouks* had left lying about without sepulture; they had thrown a very large number into pits, to avoid digging graves, and a part of the water, which had already become rare, was found to have been thus poisoned.

It would have been well to fly from the places thus infected as quickly as possible; but the care of the sick, the havoc made in the ranks of the medical men who had fallen victims of tried devotion to their duties, the necessity of organizing the service of soldier-attendants from the different regiments, and the time required for embarking the sick and for provisioning the division, did not permit us to leave for Varna before the 7th of August. The pestilence raged up to that time, but on the 9th, as soon as the column arrived upon the high plateaux of Kavarna with its air revived and made wholesome by the forests of the Balkan mountains, an amelioration in the sanitary condition was quickly felt, and the epidemic presently lost its intensity. Some days later, the division returned to its camp at Franka, where it prepared large field-hospital tents, and under the most favorable hygienic conditions. There remained but half of its late effective strength, the other half being in the hospitals, or under the soil. The loss of the *bachi-bozouks* was still more cruel; M. Cazalas estimated, that their deaths amounted to full half of their number.

The second division entered the Dobrutcha after the 1st. Upon arriving at Mangalia it found itself suddenly seized with the cholera, and struck down without mercy; but General Bosquet, in the course of his operations, observed a steady and firm policy, following faithfully the hygienic measures advised by the physicians, and rigidly enforcing them. The soldiers on the march never neglected to make their soup and coffee, however long the day's journey, or scarce the water, which was drawn from wells, few and very deep. Three hundred *arabas*, embracing the means of transportation of the 2d division, had been divided among the different corps in such a way that each had its own provisions with it, and each could watch the peasants, and their oxen, who were always ready to desert. This measure did not prevent some of the former from taking flight, but they at least left their waggons, and the beasts that drew them. The soldiers were made to supply their place as drivers. As the waggons were unloaded of their provisions by the daily consumption of the army, they placed upon them the sick, and thus greatly increased the ordinary transportation of the ambulances. At each bivouac, they dug large trenches to inter the dead. One day, General Bosquet said to an old soldier, who, pipe in mouth, was covering over his comrades with apparent indifference, "Close up this ditch, there are now enough in it." "I am in time, general, and there are more to come," replied the grave-digger, who felt himself mortally struck with the cholera. A few moments after he fell into the open ditch, and his body filled the place which he had prepared. The 2d division was not more than five miles from Varna, when the aide-de-camp of the general-in-chief came to announce that the hospitals were too full already, and that they could receive no more sick. General Bosquet replied, that he was glad of it, as he could do without hospitals, and place the sick in conditions more favorable to health. Soon after, all his sick were provided for in tents placed on the high plateaux, in the midst of the woods. The soldiers discharged cheerfully, and with a good will and singular devotion, the duties of infirmary attendants. Many recoveries attested the timely wisdom of the measures taken, and in a short time the cholera, thus treated, became almost harmless. Mother Philippon, who enjoyed great popularity among the soldiers, was distinguished among all the cantinières by her indefatigable zeal. Day and night she was on foot, and in the coarse vocabulary of the camps, in which she excelled, would ask if they were hungry or wanted anything to drink. The jokes of mother Philippon passed from mouth to mouth, and made even those laugh who had the least inclination to do so.

The second division had comparatively much less cholera than the first; and the third, not having gone beyond Bojardjik, was attacked with less severity than the second. This circumstance seemed to settle beyond doubt the fact, that the epidemic derived new energy from the fatigues which our soldiers encountered in the midst of a pestilential country.

Some physicians ascribe to conditions of dryness or humidity in certain soils, an influence upon the murderous spread of cholera. They have pushed their inquiries through the various geological series, from the granite to the tertiary formations inclusive, and have studied the modifications which the epidemic derives from them; but the facts observed almost mutually contradict each other. Some observers attribute an immunity to dry granitic regions, while others declare the same in favor of marshy districts, but the Dobrutcha has given the cruel lie to this latter theory. It has been asserted, that the cholera prevailed in the plain before we penetrated it, but this statement seems without any foundation. It is certain that the commandant of the staff, M. Balliard, who about this time visited the Danube side of Silistria, never heard the cholera spoken of, nor did the army of Omer Pasha or the village populations complain of its presence. It can no longer be doubted, that the germ of the epidemic was in some way latent in the ranks of our army, and that the slightest causes were only needed to bring it suddenly into operation.*;

If the essential causes of cholera are unknown, and if they are of a nature to escape our inquiries, the circumstances which propagate it are more and more manifest. The unhappy experiences of the Dobrutcha clearly prove that the violation of hygienic laws, insalubrity, and misery, are capable of exciting it into murderous activity, and constitute its true element. It is enough to know that the revival of this pestilence, which has so often ravaged the army of the East, has

* This generally received opinion is not entertained by M. Cazalas, according to whom the choleraic influence reigned evidently in the Dobrutcha, before the arrival of our army. "I have myself seen," said he, "cases of cholera among the native population, and the reports which I have received from these places leave no doubt of the fact. It is, however, incontestably proved from the fact that the 1st division, which suffered most, was composed of regiments from Algeria, where cholera did not prevail at the time they left; and that the *bachibozoucks*, who had never seen either France or Algeria, and who had had, so to speak, no communications with our troops, were attacked first, and most violently. Thus the choleraic influence was developed spontaneously in Turkey and the Crimea, where our troops have been much exposed to bad hygienic conditions. Such at least appear to be the facts demonstrated."

uniformly happened when the army was in a peculiarly critical position, and suffering from the influence of extraordinary depression, want, privations, and fatigues.

The specific remedy for cholera is yet to be discovered; but medical science is not powerless: it gives wise preventive measures, which are but too rarely observed, and knows the symptoms indicating the approach of the disease. An indisposition with tendency to chills, a general feeling of illness, and above all, a derangement of the bowels, with diarrhœa, are the premonitory symptoms, of which we should take careful note in times when cholera is to be apprehended.* By taking immediate care we are almost sure to escape the disease, or to reduce it to a mild form. Cases of fearful suddenness, without premonitions, are so rare, that, in the opinion of many physicians, they do not exist at all.

* All army surgeons have felt the difficulty of reaching the early stages of disease among the privates of the regiment. Habitually careless of his health, the soldier will disregard the early premonitions of an epidemic, and the sanitary precautions necessary to avoid an attack. It is only by a strict enforcement of wholesome regulations, and a constant watchfulness of the police of the camp, that men will be prevented from practices injurious to the health, and of imminent danger. Diarrhœa, as an accompaniment of cholera, or rather as a premonitory symptom, has been generally remarked. The disease is attended with a depressing effect upon the human system. In the official medical and surgical history of the British army in the Crimea, the following statements were made illustrating these facts.

"It was noticed by medical officers generally, that from the date cholera first appeared among the troops in Bulgaria, the soldier's efficiency was greatly impaired; he was no longer able to undergo the same exertion, he was fatigued by short marches, and a parade, protracted for a few hours, left him exhausted by the weight of his accoutrements, and exposure to the hot sun; and when the army arrived in the Crimea, it was evident that the physical capabilities had already been sadly reduced. In short marches—though it must be admitted generally marches very protracted—the men fell out of the ranks in great numbers; they were unable, in many instances, to carry their burdens, and their canteens, and even blankets, in some cases, were thrown away. This mysterious poison, so subtle as to traverse immeasurable distances, riding on the air, so gross as to find conveyance in moving battalions, across water, over hills and valleys, and, if necessary, through climates of opposite characteristics, was at work; and the exhausting flux which it determined, drained off the manly energies of the soldier, sapped his strength, and rendered him, in many instances, helpless as a child."—Tr.

The measures to be taken are very simple, such as placing in a bed, so as to facilitate a salutary perspiration—warm aromatic infusions—a girdle of flannel around the bowels, and attention to diet. Light attacks of the disease require no other treatment. In the cold stage it is chiefly necessary to restore the heat and circulation, and we must have recourse to warm aromatic drinks, with a few drops of ether. Rough frictions over the body, sinapisms upon the extremities, warm flannel coverings, and bottles of warm water applied to the body, are usually considered useful measures. Shampooing (massages) and friction excite the capillary circulation, and recall the heat to the cold parts, as after submersion; yet some physicians doubt their efficiency. In fact, it seems doubtful whether purely mechanical agencies can combat successfully a refrigeration caused by a depression of the vital forces. Turkish baths exert a remarkable influence, and M. Cazalas derived excellent results from their use in the hospitals of Constantinople. These remedies could not, of course, be used to any extent in the Dobrutcha. The inadequacy of the stimulant allowed the pulse and animal heat to fall almost to complete depression, and many of the sick died without any reaction.

These excitants, if pressed too far, have their danger, as they may hasten the reaction of fluxes and visceral congestions, which may prove fatal. We therefore find ourselves placed between two evils—too little and too much. The appearance of reaction is an almost infallible indication of returning health, if rightly managed. Bleeding, and acidulous drinks, modify its violence. Convalescence demands the greatest caution, as relapses are always very serious in results. It will be seen that this treatment is easy to apply; it is rational and simple, and in fact too simple to satisfy the sick, who are not always contented with common remedies.

Cholera is transmitted by the air, and is not contagious in the exact sense of the word; or otherwise, all our physicians would have been attacked. In its journeys it has two different modes of travelling: sometimes from neighborhood to neighborhood, and sometimes it advances by leaping over a district where the inhabitants appeared to be threatened, and invades, on a sudden and without notice, other places where it was least expected. In these latter cases it is probably imported; but whether imported or not, wherever causes of affinity pre-exist, no matter what sanitary precautions are taken, its visit is certain; in like manner it may disappear spontaneously without our knowing why. When circumstances favoring its appearance do not exist, it may be imported without danger; it will not develop itself. During the war of the East there was not, so to speak, a single week in which choleraic patients were not brought in steamboats to Constantinople, but yet the disease did not attack the Mahomedan population.

The excellent traditions of the army of Africa were not forgotten. They go back to the time of Marshal Bugeaud, who exhibited an example of rare solicitude for the soldier. When he commanded an expedition, he always took care in the evening to reconnoitre the route of the morrow, and learn the obstacles which were to be encountered, with the view of regulating the times of departure of the regiments, and never let the men remain a needless time with the knapsacks upon their backs. The columns started at daybreak at every season of the year, after taking coffee or eating soup; every man had a light infusion of coffee in his canteen. After marching three-quarters of an hour, he always called a halt of twenty minutes, and after that they stopped to rest only a few minutes at a time, every hour. He presided himself at the passage of the fords, requiring the men to take off, or roll up, their pantaloons, and to preserve from wet their shoes and socks. If the water was deep, they passed over single file, with ropes to serve as guards. Sentinels were placed over the springs of fresh water, found upon the road, to prevent the troops from drinking to their injury, and thus many cases of sickness were avoided. When not delayed by the firing of the enemy, the troops reached their bivouac by ten o'clock in the morning. They had then time enough to get well established, prepare a comfortable soup, wash their linen, and recover from their fatigues. The camping grounds were chosen, as far as practicable, upon elevated grounds, at a distance from swamps, and near wood and water. When wood was not expected in the next bivouac, each soldier carried a faggot upon his knapsack, and a piece of wood used as a walking stick.

The Marshal often tasted the soups of the messes, and assured himself that the flannel girdle was around the loins, and not in the knapsack. At nightfall he posted the pickets and camp guards, and to prevent thieves from stealing into the camp by night, he placed the most distant sentries along the roads, and those near the camp in the bushes, knowing that thieves abandon the regular roads when they approach the bivouac. He was the last to retire to his tent, having a company encamped close to him, so as always to have some soldiers at hand, in case of an alarm or night attack. He was annoyed whenever he saw needless fatigues imposed upon the men during a campaign, such as parades, manœuvres, dressing line, etc. Said the Duke of Isly to a young colonel who, on arriving at the bivouac, had left his men ten minutes under arms, "I see, sir, that you have never carried a knapsack upon your own shoulders." Although he admitted into his intimacy the chief physician of his field hospitals, he wished to observe for himself the sanitary condition of his men, as to their digestion, and took a glance at their alvine evacuations around the camp. He checked a tendency to disease whenever he saw it, by a day of timely repose, and a supplementary ration of rice, meat, coffee, or wine. Knowing that half of the troops that entered the hospitals, or lagged behind, ran the risk of losing their lives, and that often the men were made lame and disabled by a shoe that was too tight, he ordered the colonels never to allow a pair of shoes to be given out, until they had been carefully tried on: and to frequently make themselves sure of the condition of the shoes themselves, and see that they were daily softened by a coat of grease. Following the example of Marshal Clausel, he assigned a company of men to the service of the field hospitals, to prepare the tents, and provide wood and water. He often visited the sick and wounded, and his presence restored their morale. For all these cares he was called the father of the soldiers, and he always found his troops, in time of combat, energetic and full of health, enthusiasm, ardor, and warlike courage. He could ask of them the disarming of the most ferocious tribes, the only means, in the eyes of the illustrious Marshal, of securing their submission, and the empire to our arms.

The painful impression made by the expedition of the Dobrutcha, was rapidly effaced. The allied armies embarked for the Crimea, to enter upon the real campaign, in which battles and new diseases soon rendered the establishment of many new hospitals necessary. From 1854 to 1856, nineteen French hospitals were successively established at Constantinople, in buildings of four different kinds: permanent barracks, Turkish hospitals, palaces, and wooden barracks. To recall the circumstances which caused the establishment of these hospitals would be to record the most melancholy epochs of the campaign.

CHAPTER II.

THE HOSPITALS AT CONSTANTINOPLE.

On the 14th of September, 1854, the three allied fleets disembarked, without resistance, upon the soil of the Crimea, at Oldfort, 137 cannon and 61,200 men; of whom 27,000 were French, an equal number English, and 6,000 Turks. The want of good drinking water, which would have to be brought a distance of three miles and a half, compelled the allied armies to use a brackish water, obtained by digging pits along the sea shore; but this became so bitter when it was boiled that they abandoned it in making soups. They were also without wood. The consequences were, derangements in the digestive organs, which brought on some of the symptoms of cholera, especially in the English army. Our allies, having been unable to bring their large tents, which were inconvenient to carry, were furthermore exposed by night to dampness and the abundant rains, while our African army had with them their little shelter-tents.

On the 19th an order to advance was hailed with a shout of enthusiasm. The divisions were doubly happy in marching upon the enemy, and in leaving these bivouacs, destitute as they were of wood and water. We know with what vigor and success the Russians were attacked, on the morrow, in their formidable posts at Alma; upon their left, by the troops under General Bosquet, whom nothing checked, neither canister nor the ruggedness of rocks; upon their right, by a portion of the valiant English army; and in the centre by the bayonet charges of the troops of Generals Canrobert and Napoleon.

The transportation of the wounded upon *cacolets*, and litters borne upon the backs of mules, again demonstrates that this system is better than ambulances upon wheels.* Whenever a man fell wounded in the ranks his comrades carried him a few paces in the rear, where the muleteers took charge of him. The battalion surgeon, or those who attended the ambulance, applied the first dressings and made him lie down in a litter, or sit in a kind of chair called the *cacolet*, in which he was then carried to the field hospitals, placed as well as might be out of the reach of the bullets. There each officer and soldier waited his turn to be operated upon, except that those more severely wounded took precedence.

The victory achieved on the 20th of September, poured into our field hospitals 1,033 wounded French

---

* In 1835, in the expedition of Mascara, in the province of Oran, I realized the idea of carrying litters upon camels, one upon each side. This method has since been perfected. The camel, in walking, has a balancing motion, and the wounded were placed rather high. We have exchanged, with advantage, the camel for the mule, an animal more gentle and more easily managed.—AUTHOR'S NOTE.

The *Cacolet* consists of a pair of chairs with strong side arms and concave backs, which hook together and hang across a mule. The seats are attached to the backs by hinges and straps, so as to be adjusted to the most convenient angle for the patient.—TR.

and several hundred Russians. Both were, immediately after the first dressings, carried on board the fleet, and thence to Constantinople, where, on the 24th of September, was established the hospital of Dolma-Baktché, located about one hundred rods from the Bosphorus. This hospital, reserved almost entirely for the wounded, embraced two isolated and entirely distinct buildings; the one more elevated was the artillery hospital of the Ottoman Guard, and the other, about twenty rods below, contained 600 beds. The ships arriving from the Crimea, anchored at the entrance of the Golden Horn near Top-Hana. The wounded, placed upon tenders and landed at Dolma-Baktché, were carried upon litters by attendants or Turkish soldiers. From September 24th, 1854, to April 1st, 1856, this hospital received 8,582 invalids, mostly wounded; and of these 2,318 died. The management of this important service was intrusted to a skilful director, Dr. Salleron.

The officers wounded at Alma, caused the establishment of the hospital of Canlidjé upon the Asiatic side and directly adjacent to the Bosphorus. The viceroy of Egypt had liberally placed at our disposal this domain, which was his country residence. Beautiful undulating gardens, a pure atmosphere, and elegant kiosques, made this an enchanting place of sojourn. Near us was the palace of Fuad-Pacha, minister of foreign affairs. Two young Armenian girls, belonging to his harem, eluded the vigilance of the eunuchs. Their songs and playing upon the piano had attracted the notice of two aides-major, whom they saw through the lattice of their windows. They were smitten with their admirers, and one fine day escaped in the costume of one of the pacha's sons. On the next day they were taken back to their prison. This escapade would have assumed the proportions of a grave event, had it not been for the prudence of Fuad-Pacha, who contented himself with recovering the fugitives. It was, nevertheless, asserted—wrongly, I have no doubt—that according to the Turkish custom, these two unfortunates were put into a sack and cast into the Bosphorus. Subsequently the wounded officers exchanged the hospital of Canlidjé for the house of the

Russian ambassador. Two or three hundred beds, placed in the palace of Mehemmed-Ali, were appropriated to the soldiers.

It will be remembered that the allied armies landed in the Crimea with nothing but light artillery, which was quite unable to contend against the heavy marine artillery which bombarded them from the roadstead of Sebastopol. It was necessary to prepare for a regular siege. The works of investment and circumvallation were begun on the 9th of October, and rapidly pushed forward; numerous battalions and companies of sharpshooters protected the laborers. Night and day one half of the army were exposed to the iron storm and the vicissitudes of the season, while the other half snatched a short repose in order to take its turn at the work. New troops arriving daily increased the number of the sick, while the failure of the fire, opened October 17th, 1854, against the place, by the vessels of the two combined fleets, and by batteries mounting 126 siege guns, added new numbers to our wounded, and seemed to indicate that the city of Sebastopol, then defended by a garrison of 32,000 men,* and by the army of reserve under the orders of Prince Menchikoff, could not be taken by a sudden dash. The remittances of sick and wounded from the Crimea to Constantinople, followed one another rapidly. In October two important hospitals were opened, one for 1,200 sick at Ramis-Tchiflik, a fine barrack situated upon the plains of Daoud-Pacha; and the other upon the heights which command the Bosphorus towards Pera, in the buildings of the preparatory school, which were arranged to receive 400 beds. In the following months new hospitals were opened. In the extensive gardens of the point of the old seraglio, at Gulhané, the military engineers had erected barracks for 1,800 sick; above that, the University palace, a fine edifice of cut stone, and still unfinished, was prepared as a hospital for 1,400 beds.

These two establishments, opened in the old aristocratic Ottoman quarter, in the very heart of Stamboul,

* Of whom 21,000 were marines, rendered available in the sinking of the vessels to serve as a barrier to the roadstead.

denoted to what degree of tolerance the Turks had attained concerning us. In the suburb of Péra, there were added to the hospitals already established there, that of the Military School, containing one thousand one hundred beds, reduced soon after to five hundred by a fire, and that of the Parade Ground, containing twelve hundred beds under shelters. The barrack of Daoud-Pacha, designed as a place for convalescents, became itself a hospital for one thousand two hundred sick. With such vast resources for hospital accommodation yet barrack-camps for twenty-five thousand men were being constructed at Muslak, on the high plateaux which border the shores of the Bosphorus, and proved of incalculable service in time of the typhus. The central pharmaceutical department, intrusted to supply the medical service of the Crimea, and of Constantinople, was established upon the sea-shore, near Bachisstach, in the palace of a pacha. The Minister of War had sent to the army of the East, new and complete material for thirty-five hospital establishments of five hundred beds each. I have spoken before of the difficulties encountered in the Crimea, in taking care of the materials, and in washing the bed clothing and the linen. We were surprised to find these difficulties at Constantinople. To surmount them, it was only necessary to establish small steam wash-houses, like those which some military hospitals possess, as, for example, that of Nancy. For this purpose, a movable furnace containing a steam generator is necessary. The steam entering at the base of the double-bottomed coppers, acts mechanically upon the linen, impregnated with an alkaline solution, and hastens the formation of soap-suds. Without going into further details of the operation of this economical apparatus, easy to set up and transport, I would add, that it only requires *four hours* to empty a copper of three and a quarter tons, and eight hours by the method of Thiéry, to empty four coppers at once. How useful these washing machines would have been, not only in our hospitals, but in our field-hospitals, and even in our stationary camps.

About two-thirds of the fever patients received into

the hospitals of Constantinople, were attacked by diarrhœa or dysentery. The diarrhœa was so common, that we might say that sickness was almost always preceded by a diarrhœa in the acute stage, and terminated by a diarrhœa in the chronic stage. This sad complication was not peculiar to the army of the East, but has been observed in all armies in campaign ; it is concomitant with the soldier's life, to the poor rations, home-sickness, and the thousand influences which it is not always possible to foresee. Dysentery has almost always for its first symptom, a more or less active diarrhœa, of which it is, as it were, the second stage. The intestinal changes proceeding even to ulceration, indicate alike the anatomical lesion in the two diseases.

Acute diarrhœa, so common among soldiers upon beginning a campaign, is often cured in a few days by rest, regimen, the wearing of a flannel girdle, and if needed a few drops of laudanum.* If it were possible to treat it always by these simple means, and to prevent relapse by hygienic care, we might certainly reduce by more than half the number of cases of disease, and of death. An emetic, or an emetico-cathartic, will almost always drive off in a short the gastric difficulties with which this disease becomes complicated. In the chronic or advanced stage, astringents, which are so much extolled, only give an ephemeral rather than a real relief, and are often more pernicious than useful. The best tonic is good wine, in small doses often repeated, the effect of which should be carefully watched by the physician.

* Feculent, gummy, and sweetened drinks, are almost always given in intestinal disorders, as emollients. In a very interesting work upon antiphlogistic and emollient medication, by Prof. S. Delioux de Rochefort, he has examined the action of these so called emollients. He concludes that, feculent, gummy, and sweetened drinks, being soon decomposed and absorbed upon entering the intestines, their emollient topical application is null, unless we consider them as nutritious drinks, perhaps useful in so far as they make agreeable change in the diet of the sick. In a word, they are aliments, not medicines. These views appear to justify the practice of the English physicians, who are as sparing of diet-drinks as we are prodigal of them in France. It is true that our sick believe they cannot be cured unless they take infusions of all kinds, which harm oftener than they benefit.

A light dose of opium alone, or better still, given with ipecacuanha, or sulphate of magnesia in small doses, has been more efficacious than all other therapeutic agents. A severe regimen can alone prevent a relapse, which is very often fatal.*

This disorder would have committed greater ravages, had it not been for the moral energy which through the whole campaign sustained the French troops, and which never showed itself so strongly as in the most critical moments.

---

* Diseases of the bowels, the constant scourge of armies, affected the English troops to an alarming extent in the Crimean war. For one-third of the eighteen months embraced, they presented a more fatal aspect than had ever before been recorded. The exciting causes might be divided into two great classes.—First, an epidemic or choleraic constitution of the air, with the seasoning agencies of a hot climate, and 2d, the hardships and privations of a winter siege, in an inhospitable climate. The mortality was chiefly to be ascribed to the latter class of causes. From April, 1854, to June, 1856, inclusive, 55,767 cases of diseases of the stomach and bowels were admitted into the English hospitals, of which 7,611 were from the cavalry, 7,014 from the ordnance, and 41,140 from the foot guards and infantry. Of these, 5,950 were fatal; 281 from the cavalry, 487 from the ordnance, and 5,182 from the foot guards and infantry.

The scorbutic taint gave peculiar virulence to this class of diseases, and the prevalence of ulceration of the intestines, especially towards the lower part, was perhaps the most constant pathological condition observed. Surgeon Macleod remarks upon this condition as follows:

"The immense majority of those who served during the early part of the war were so affected, the ulceration being rather of recent than ancient date: and this remark does not apply to those who died of abdominal affections, but also to those who succumbed from other diseases or from wounds. It is also a fact which I have had many opportunities of verifying, that men killed in action at a time when they were apparently in the possession of health, or rather, as it should be put, men dying shortly after receiving severe wounds when seemingly robust, were found to have ulcers in their intestines, sometimes of a very extensive character. To this it was not uncommon to find diseased kidneys and lungs added. The disease, in these cases, might not be active at the period of death, but it was ready to break out whenever any injury or operation made an extra demand on the powers of life. It is of importance to note this extraordinary prevalence of undeveloped disease—this deceptive character in the appearance of the men—as bearing on their behavior under accident."—Tr.

The importance of the Malakoff bastion was fully recognised, and preparations for its attack were pressed with vigor. The Russians had executed with rapidity a series of works and counter-approaches, which it was resolved to take on the night of February 23–24, 1855. General Bosquet passed through the trenches where the soldiers were half leg deep in mud. He was preparing his troops for the combat, when he met a sentinel, who had just before been wounded in the head, and who presented arms. Seeing the blood flowing from the wound, the general asked him why he did not go to the field hospital. " My shoes are full of holes," replied he, alluding to the avidity with which his comrades disputed for some of the Russian spoils ; " there is to be a distribution of boots to-night, and I want to be there."

It was not cholera and dysentery alone that peopled our hospitals of the East, but fevers of various kinds. The miasms exhaled from the putrid decomposition of vegetable matter taint the air, and produce in the economy the same effects as poison, which nature seemed to seek to rid herself of by critical and periodical fevers. This special fever, which we have named intermittent, to distinguish it from continued fever determined by other causes, is characterized by three well marked periods; first the cold stage, then the hot, and lastly the sweating stage. However, it rarely showed itself in the Crimea in this pure and distinct form; but commonly the accession was incomplete—the heat would steal on without the initial chill, and without being followed by the sweating. This disease appeared to be no other than a modification of continued fever. It was very rare to find diseases, of themselves continuous, run through all their phases, without becoming complicated with intermittent phenomena. The fevers were then generally remittent. They are thus called when they are composed of a continuous febrile element and of an intermittent febrile element.*

---

* The tendency of malaria to give a remittent form to all diseases that may appear in a given district where this influence prevails has been often noticed.

The Russian physicians regarded remittent fever as endemic in the Crimea, where the Tartars contracted it as readily as strangers. This malady has been considerably developed in their army, and they principally attribute it to the insalubrity of their subterranean huts. Among our soldiers the accessions of remittent fever were rarely complete. The slightest intermittent attack proved rapidly fatal when it appeared in the course of a continued fever, for the economy had then already received serious damage from privations, diarrhœa, scurvy, and other chronic diseases. In these complicated cases, in which the intermittent was only secondary, the first care was to attack the continued febrile element by emetics if there appeared a gastric difficulty, by bleedings if symptoms of plethora were noticed, &c. At the same time, upon the first appearance of intermission or remission, it was necessary to hasten or prevent their return by two or three doses of quinine,* of from fifteen to thirty grains. The attacks of pernicious intermittent

The surgeons of the American army in winter quarters near Washington, in the winter of 1861–2, observed this tendency in nearly all their diseases, and derived a corresponding benefit from the use of quinine.

The whole of the region occupied by the army of the Potomac was more or less malarious, and autumnal intermittents are familiar to the civil practitioner. In some of the more sickly of the regiments, the low condition of health might fairly be traced to the depressing effects of summer camps in low valleys, which left the men with constitutions impaired, and liable to the invasion of diseases. These usually assumed a typhoid form. The percentage of sickness to the effective strength varied from one-third of one per cent. to eighteen per cent. in a regiment, the average being about five and a half. Some of the sicklier regiments were from the Northern border of the Union; and it was observed that soldiers enlisted in the naval districts, although often a rough and apparently hardy class of men, endured the open winter and humid climate of the Potomac region with less resistance than those from cities.

The army of the Potomac, abundantly fed and clothed, presented a favorable comparison with the British army in time of peace, in which six and a half per cent. of the force otherwise available are reported "in hospital." In the Peninsular War under the Duke of Wellington, in 1808–14, it was twenty-one per cent., and their sick list varied from nine to thirty-three per cent. at different periods.—TR.

* The sulphate of quinine is always to be understood by this term when used in this work.—TR.

fever are caused by a deep paludal intoxication. In the Dobrutcha a certain number of cases were observed, but it was rare in the Crimea.

The still increasing numbers of our fever patients rendered the hospitals at Constantinople insufficient for their accommodation. The Sultan, with an unprompted generosity, offered a palace then scarcely finished, which bore his name, and became thenceforth the hospital of Pera. The architecture of this palace was in the oriental style, very handsome, and of great solidity. Each angle had a tower which supported another story, and a central minaret, adorned with many ranges of balconies, boldly pierced the clear azure of the heavens, and lent to this monument an aerial appearance, without detracting from its majestic character. The ground floor, elevated about six and a half feet above the earth, and the first story, presented broad and immense galleries, lighted on the side of the court by arched openings, the arches resting upon elegant pillars. The openings were closed by large windows. These galleries were intended to serve exclusively for promenades in bad weather, and it is to be regretted that the necessity of lodging 2,000 sick required us to employ this space. We have already noticed the danger of collecting large numbers of men afflicted with severe maladies and forced to remain all the time in bed. It is offering a premium for infection, which is the cause of the greatest number of deaths. Abundant water, of good quality, was drawn from the fine forests of Belgrade, where visitors admire the gigantic aqueducts of Constantine, and the still more wonderful sluices, which are held in place by enormous blocks of marble, brought thither at great cost in the reign of the Sultan Mahmoud. In front of the hospital of Pera rise the shores of Asia, the city of Scutari, and its deeply ravined knoll, which descends to the Bosphorus and commands the great Field of the Dead, planted with evergreen trees. The sick arriving from the Crimea were debarked at Bachistach, distant about a mile and a quarter from the hospital, but the ascent was so steep that convalescents themselves could scarcely make the journey on foot. At the head of this great establishment, the renowned physicians, Messrs. Scoutetten, Morgues, and Cambay, were successively placed. Each of them sought to reduce the figure of hospital inmates, but still the beds were always full. The Crimea sent us new sick daily; every ship brought two or three hundred. It may give some idea of the importance of the hospital of Pera by mentioning that it received 27,500 sick during the twenty-two months of its existence, of whom 9,460 left entirely cured, 13,000 were sent to France or turned into other hospitals, and 5,040 died.

After the taking of the Malakoff bastion, the hospital received 800 sick in one day, of whom 595 were severely wounded Russian prisoners. The greater part of the latter would not at first consent to the grave operations which their condition demanded; it was only later, upon seeing their comrades die, that they consented. Unhappily, this delay was fatal; but, nevertheless, they survived in greater numbers than our own soldiers, because their constitutions were less deeply impaired by fatigue and privations. Moreover, a rough and chiefly manual education, strict sobriety, and a coarse but abundant aliment, caused a great preponderance of the muscular system* in the physique of the Russian soldier; the nervous element being much less developed, they suffer less in operations than our own soldiers. The progress of their wounds is more rapid and regular, the febrile reaction less marked; their power of assimilating food greater, and requiring greater quantities; their blood appeared richer and more plastic, and their arterial system more fully developed—so much so, that after each operation it was necessary to tie more arteries than in the wounded of our own army. They evinced strong tokens of gratitude towards the French surgeons, who

---

* The experience of the army surgeons of the United States among our Western Indians has led to a somewhat similar remark. The reparative forces of nature appear to be more powerful than with the whites, and they will survive a greater bodily mutilation. This has been ascribed by Dr. Day to the greater simplicity of their diet, and may be in part due to a less delicate susceptibility and the greater development of the muscular as compared with the nervous system.—TR.

bestowed the same care upon them as upon our own soldiers, among whom they lodged. None sought to escape. Our rations of white bread, more easy of digestion than the Russian army bread, proved insufficient, and had to be increased. These soldiers wore pictures of the saints, or copper crosses, suspended around the neck by a tape; they recited daily their prayers in their beds, taking no notice of the persons in the room.

From the 21st of May, 1853, the house of the Russian ambassador at Pera remained closed. On that date Prince Menchikoff hastily took his departure. Whilst the French and Russian officers and soldiers encumbered our hospitals at Constantinople, the doors of the palace remained inexorably closed, although it was of size sufficient to receive 400 sick. A force of 30,000 men, 22,000 of whom were of the guard, were assembled in the fields of Maslak, to reinforce the army in the Crimea. They experienced many attacks of cholera, which threw many new sick upon our hospitals. At length, after the numerous and bloody conflicts of April and of May 1st, 1855 —after the grand and terrible artillery duel, which gained us some important works of defence, it was decided to lodge the French and Russian wounded officers in the house of the ambassador. All the furniture was carefully carried into rooms reserved for its storage. M. Lelouis, the physician in chief, a man of unquestionable merit, took charge of the wounded, and devoted himself to them, but still this hospital soon showed signs of infection. The wounded became affected with hospital gangrene, and afterwards the typhus, imported from the Crimea, was propagated from bed to bed. When peace was signed, the French government expended large sums in putting this palace in good condition, and made it even finer than before, by repairing injuries of older date.

At the beginning of the campaign, the Sisters of Charity had opened, in the suburb of Pera, a hospital, which soon became highly sought for by the officers. Each invalid received, in a room by himself, the most tender and intelligent care, and could enjoy the services of any military surgeon he should choose. This convenience was very highly appreciated, and the hospital of the Sisters was always full.

Among the French soldiers received into the hospitals at Pera, there were many who had been wounded in the frequent street quarrels of this suburb, whose mixed population, so different from that of the Musulman quarter of the Stamboul, embraced a great number of renegades from justice, of every nation. At Pera, crimes were committed with impunity in open day; assassinations occurred in the middle of the street, and the passer-by went on his way, as if he had seen nothing. By request of the general of division, Larchey, superior commandant at Constantinople, the French ambassador, M. Thouvenel, obtained permission to establish a French police at Pera, and our armed police guards rendered the most essential service. They arrested the malefactors; but here a new difficulty was encountered, for these miserable wretches were reclaimed by the consuls of their country, who, under the pretext of trying them, allowed them their liberty. It generally ended by granting them a hearing, and by their giving a sufficient security for good behavior.

At the time when the house of the Russian ambassador at Pera was turned into a hospital, the most bloody conflicts of the campaign had occurred, and some months after, the taking of Sebastopol seemed to put an end to the Crimean war; but the task of our medical staff was far from being ended, and if the numbers of our wounded were less, those of the sick were largely increased, under the double influence of scurvy and typhus, and in proportions which excited the greatest solicitude. Then, as at the opening of the campaign, the Turkish government showed itself most kindly disposed to second the efforts of the French administration. Contrary to all antecedents, the Sultan had attended a sumptuous ball given by the French ambassador. The French and Ottoman troops fraternally formed the double lines upon his passage; salvos of artillery announced his entrance into the palace of the ambassador. Abdul-Medjid was conducted into an elegant saloon, reserved for his use, where I had the honor of being presented to him. He spoke to me with lively interest, of the body of Turkish troops which I had visited at Eupatoria, of the health of his soldiers and ours, and pressed me to visit the Turkish military hospital at Constantinople, concerning which he desired my opinion. The Sultan understood French, and spoke it in its purity, but with a timid reserve, so that his minister of foreign affairs, Fuad-Pacha, who had made medical studies in Paris, hastened to supply the words to his thought when he hesitated. His countenance, naturally somewhat sad and thoughtful, was animated in conversation, and showed at times a remarkable expression of ingenuousness and benevolence. He made his entrance into the ball, in the midst of all the high functionaries of his empire, covered with embroideries in gold, and crosses in brilliants. His costume was of a rich plainness: a cap of red felt without ornament, a little black cloak with upright collar, sparkling with large diamonds, a European coat, and the broad band of the Legion of Honor. The old Turk party was greatly troubled on this occasion; in their alarm they thought that the Sultan, in accepting the broad band of the Legion of Honor, had been converted to Christianity; but to quiet them, they were made to understand that the star of the Legion of Honor was composed of five points, and not of four, as in the Christian emblem.

The Sultan advanced with slow and measured steps into the ball-room, looking to the right and left with a tranquil and easy air, although it was the first time he had been in such an assembly. He took his place upon a reserved seat, where he appeared to follow with some interest the movements of the dance. I know not what impression Abdul-Medjid received from this array of beautiful women and brilliant toilets, but I doubt if it was favorable to female emancipation in Turkey. He retired at an early hour, with the same ceremonial. I remarked that the assistants kept some distance from his person, and I learn that this is not simply from respect, but on account of the aversion which contact with a man inspires, and which is explained by remembering the disastrous epidemic so common in the East. The Sultan lays off, never to resume, the garments which a man has touched. We know that he is served exclusively by the females of his harem. He never addresses a word to any one in public; but once or twice, to the great astonishment of Musulmans, he laid aside this traditional habit, in favor of General Larchey. He fixes his looks, for a longer or shorter time, upon the person whom he meets, according to the degree of esteem which he wishes to evince. In this mute language of the *padishah*, are foreshadowed the inner sentiments which words fail to express. I was able to read it perfectly, during the filing of the high functionaries of the empire before the Sultan, on the day of the ceremony of the *beiram*, or kissing the feet. The filing past continued more than an hour, during which the looks of Abdul-Medjid did not rest upon more than twenty persons. I observed that they made only a feint of kissing the feet, and each time that an indiscreet worshipper touched the Sultan, a slight motion testified to the keen and disagreeable sensation which had troubled the reverie of his sovereign.

CHAPTER III.

THE SCURVY AND TYPHUS.

THE leisure which marked for us the beginning of winter in 1856, was short. The attention of the medical corps was soon called, as I have said, to two grave epidemics—scurvy and the typhus, which came upon us with cruel intensity.

In the Crimea as elsewhere, the scurvy was induced by debilitating causes, a diet too uniform, composed often of salted meat, and a slender supply of fresh vegetables, personal filthiness, fatigue, home-sickness, putrid emanations, and above all a cold, damp, and rigorous winter. The first stage of the scurvy is marked by a change in the constitution of the blood, but without any very apparent external or local symptoms. A general tendency to hæmorrhages, great muscular lassitude, deep-seated pains, especially towards the feet (pains which physicians have mistaken for a specific disease, calling it *acrodynia**), a slackening of the pulse, a loss of appetite, a marked discoloration of the skin, and a remarkable dilatation of the pupils, these are the symptoms of the first stage of this disease. Soldiers were rarely sent to the hospital during this stage, but almost all of those received for other diseases, had at the same time scurvy in the first stage. In the second stage, the gums swelled, softened, and ulcerated, emitting an infectious and abominable odor. A sister of charity died of gangrenous inflammation of the pharynx, from having inhaled the breath of a scorbutic patient, while touching the ulcerated gums with a pencil wet in hydrochloric acid. The teeth become loose, and more prominent; the lower extremities infiltrated, present livid spots, extensive bloody effusions, especially in the internal parts, extensive serous engorgement. The muscles, losing their elasticity, become hard, like wood, and the patient can no longer walk. In the third stage, the greyish ulcers of the gums extend to other parts of the mouth, sometimes perforating the cheeks, in the form of gangrenous spots, of which the parotid glands were the principal seat. They entirely consumed the tonsils, and caused caries of the maxillary bones. Hæmorrhages occurred from the mouth, nose, urinary passages, and the intestines, the pulse became extremely feeble, and the falling away and softening of the tissues continued, until the serous scorbutic cachexy usually

---

* From the Greek words ἄκρος extremity, and ὀδύνη pain. This name was applied to an epidemic disease which prevailed in Paris in 1828 and 1829, and of which the prominent symptoms were a painful formication usually felt in the feet and hands.—AUTHOR'S NOTE.

This affection, like the "burning feet" noticed among the sepoys of India, has been by some writers looked upon as a sequel of rheumatism. The Parisian disease was described as attended with formication of the feet and hands, streaking along the edges, a varying degree of sensibility, an œdema, dark patches on the limbs, desquamation of the epidermis, lancinating pains, and great heat on the parts at night. In severe cases, delirium, inflammation, paralysis, marasmus, and death followed.—TR.

ended by distinct asphyxia, following the swelling of the glottis and epiglottis, preventing the air from reaching the lungs. Often congestions form in the viscera, which were found after death, infiltrated with discolored and very impoverished blood.

Scurvy prevails as an epidemic, and is rarely found without being complicated with chronic diarrhœa, intermittent or remittent fever, bronchitis, pneumonia, or other diseases, which more directly cause the mortality attributed to scurvy. The treatment throughout, should be hygienic, rather than therapeutic. In leaving the Crimea, the scorbutics escaped the predisposing causes. At Constantinople, and especially in France, a diet of fresh vegetables, timely adopted, almost always sufficed to operate a cure, when the disease was simple, and without complications.

The English army escaped scurvy in 1856, from which it had suffered so greatly the preceding year.* As to the Russians, they were, according to their army surgeons, but lightly affected by it, and they attribute this almost complete immunity to the daily distribution of cabbage and fresh meat. When the war was declared, Russian families had sent from every point of the empire, under the name of a national gift, *arabas* loaded with wheat, and drawn by oxen. This patriotism relieved and powerfully seconded the efforts of the Czar for provisioning his armies.

* The English suffered less from the scurvy in well defined forms, than from the scorbutic taint, which was wide-spread and complicated with other diseases, especially fever, and affections of the bowels. For the first six months of the siege, all morbid action in the older residents was more or less distinctly marked by scorbutic symptoms. It was much more prevalent among the infantry than in the ordnance and cavalry branches of the service. This disproportion was observed chiefly in the winter of 1857, when the privations and sufferings of that arm were greater than in the others. All the facts observed, only confirmed the truths previously established, that defective innutritious diet, improper shelter and accommodation, overcrowding and filth, as distinct from excessive labor and fatigue, were the essential causes of this disease. The English, in the Crimea, had 1,763 cases of scurvy admitted from the infantry, of whom 172 died. From the Ordnance 105, and from the Cavalry 228 were received, and from each of these, three died.—TR.

F. The Ottoman troops encamped at Eupatoria, sent monthly to Varna a thousand scorbutic patients, mostly those gravely affected. A short sojourn in a place where fresh vegetables abounded, re-established their health. To apply this sovereign remedy to the scorbutics of our army, it was only necessary to find some neighboring island in the Archipelago, and to obtain permission to occupy it. Mytilene appeared to unite the necessary conditions, and on the 1st day of December, 1855, I went thither with Messrs. DeCourville, a captain of engineers, and Quesnoy, major physician, upon the steamer *L'Ajaccio,* attached solely to the personal service of the French ambassador, and which M. Thouvenel placed at my command. M. Laurent, the captain of the vessel, notwithstanding the foul weather, brought us to the island in thirty-six hours. The French consul, M. Didier, procured us horses led by *cavas,* or couriers. These cavas follow the rider, and are never far from him, whatever may be the pace of the horse. I was pained at seeing my cavas run by my side over the rocky roads, and hoping to leave him behind, rushed on at full speed, but was surprised to see him arrive before me, ready to hold the stirrup, and to assist me in alighting.

Mytilene, one of the largest islands of the Archipelago, is the ancient Lesbos, so famous for its wines and its courtesans. It is about half way between Smyrna and the Dardanelles, triangular in form, the angles terminate by Cape Mativa on the north, Cape Sigri on the west, and Cape St. Mary on the east, and its circumference is about one hundred and twenty miles, eighteen miles in length by thirty-six in breadth. The soil is very sloping, and free from marshes. The highest mountains are on the west side of the island. Mount Ordinus may be seen at a distance from forty-five to sixty miles; and Mount St. Elias, at the eastern end, and south side, forms a high plateau, crowned by Mount Olympus, having a height of three thousand and eighty feet. Besides sundry anchorages, the island has three good harbors on the southern side, of which Port Langan is the largest, Port Sigri, and lastly Port Olivier, one of the most important in the Archipelago. Port Olivier is only five miles from the city of Mytilene, and extends eighteen miles into the land, with a breadth of nearly four miles. It is entirely shut in by high mountains, from the violence of the winds. The olive trees with which the heights are crowned, give a magnificent effect, and from these the port derives its name. This port might easily contain a fleet of a hundred vessels. It must be entered by a south wind, and can only be left when the wind is north; but steam tugs would relieve this difficulty. The mountains on the west are covered with pines and fir-trees of large size, whose timber furnishes materials to the ship-yards for construction of large vessels. A dozen pretty villages are perched on the gentle swell of the mountains. At the head of the bay, there exists an establishment of thermal waters, slightly saline, and of 86° F., called Quindros, having two large marble pools, sufficient to accommodate a hundred bathers. These waters, which enjoy a great celebrity in that region, could be used by our sick; they would doubtless prove very beneficial in the indurations and pains in the limbs left by the scurvy.

The Sultan collects one tenth the value of all the products of the island. Mytilene, in 1850, exported three hundred thousand quintals of olive oil, but the severe winter of 1851 injured the trees, so that the production was temporarily reduced to one hundred thousand quintals. The island has numerous mulberry plantations, and exports annually about twenty-two thousand pounds of silk. Its production of wheat is insufficient for the islanders; it has many sheep, and their meat, which is excellent, sells at about five and a half cents the pound, and the wool at five and a quarter cents the pound. Cattle are kept for labor; and those used for food, are imported from Asia, whose coast is but ten miles distant. The horses are very small, and resemble those of Corsica. Cow's milk is rare, but that of the goat is abundant during ten months of the year, and from this they make excellent cheese. Fresh vegetables are abundant, and of very low price. I have seen cabbages sold for a cent, which in the Crimea would have cost fifty cents. The potatoes are of good quality, oranges and lemons abound, and in the fish markets, mullets, dorads, and lobsters were plenty. The wine is warm, generous, and aromatized with mints, but this in my opinion weakens its quality. It is said, that rich mines of antimony are in prospect of being opened; fine quarries of marble, and even of mineral coal, discovered at Policnity, have not as yet been worked.

The population is estimated at 70,000 souls, of whom 20,000 are Turks, and of these 10,000 to 12,000 live in the city. The remainder of the inhabitants, almost all of Greek origin, are scattered among 74 well-built villages, and appear in easy circumstances. The climate is very healthy, soft, and temperate, and the orange-tree thrives in the open fields. Diseases are rare, and intermittent fever is, so to speak, unknown. The people live to an advanced age. Water is abundant, and of excellent quality. Mytilene is so celebrated for its great salubrity, that many sick from other islands in the Archipelago go thither to pass the time of their convalescence.

A hospital for convalescents would be happily placed in this favored region. The city of Mytilene is commanded by a large citadel. This citadel, built by the Genoese, of hewn stone, stands out like a promontory, and its ranges of batteries rise to an elevation of about 260 feet above the level of the sea, from which it seems to rise as a single mass. This fortress includes numerous magazines, some empty, others containing old cannon frames. It is occupied by only 400 native troops. It would be easy to arrange for the use of the sick, some of the magazines and uninhabited houses which the Turks had built in 1820, as a means of security in the time of the Greek war for independence—so as to accommodate 300 convalescents. There were still other buildings that might be taken. Upon the summit, 110 yards in the rear of the citadel, was a Turkish barrack, which only needed the walls to be whitewashed, and the number of windows increased. West of the city, in the midst of fine culinary gardens, stood the school of the Greek community, with very spacious and convenient rooms. The *conak* Moharem-Bey, and the Metaxa house, were two vast Turkish palaces, immediately available, and in perfect preservation. The Pacha himself offered me the palace of the former governor, but this was falling into ruins, and could not be safely inhabited. He also tendered to me his country residence, situated about six and a quarter miles south of the city, on the sea shore, and adjacent to a small landing. In riding there, along the shore, I passed through a beautiful grove of olives, in the midst of which was a cluster of pretty villas. In short, my visit to the different establishments on the island adapted for hospital use, convinced me that it would be easy at once to lodge 785 convalescents at Mytilene, in five buildings, isolated, it is true, but grouped within a distance of 500 or 600 yards. This subdivision, doubtless, rendered impossible the erection of a hospital of convalescents, is such as required by classic traditions, but these traditions do not appear to me imperative when it concerns convalescents, to whom liberty of movement and of walking in the open air, are necessary. It suffices to create depots of convalescents, organized and living like the companies of a regiment. There might also be prepared, at small expense, two establishments under tents, each to contain 2,000 scorbutic patients, one at the country-house of the Bey, and the other near the thermal springs of Quindnos.

A learned physician settled on this island, M. Bargigli, afforded in this examination very marked and valuable assistance. The governor of Mytilene, Ismaël Pacha, said to me, "Hasten, for the English have already sent a commission to explore the island, and doubtless they will not delay coming;" and he kindly added, "I would rather have the French here than the English." M. Thouvenel, on his part, procured permission of the Sultan to put our plans immediately into effect. M. Darricau, the intendant general, Councillor of State, and Director of the War Department, wrote to me on the 29th of December, 1855, as follows:—" I have read with lively interest your report upon the island of Mytilene. Your account shows perfectly to what extent we may there form an establishment. The Minister has addressed to General Larchey and Intendant Angot, instructions to decide them on this point. There, our soldiers, under a pure sky, and in the midst of the comforts which you have described, may be able rapidly to recover their wasted strength." Unfortunately, delays and difficulties after the peace was signed, prevented a hospital and the encampments from being formed at Mytilene, where thousands of our scorbutic patients might have rapidly recovered their health, and not have furnished at a later period so large a contingent to the typhus.

I come now to speak of the second and most terrible of the epidemics, with which we contended in 1856. There has long been known a disease which develops itself specially among people crowded together into close quarters, and subjected to miasmatic influences. It is called the fever of camps, jails, ships, and hospitals, Hungary fever, Naples fever, and the contagious typhus of Mayence. Its principal features are—stupor, with delirium, an eruption upon the surface, and the power of transmission from the sick to a person in sound and perfect health. The appearances, during thirty years, of this disease, in the duchy of Posen, at Reims, at Philadelphia, at Edinburgh, at the convict quarters at Toulon, and in the prisons at Strasbourg, have happily been too short and too confined to allow us to observe the differences which distinguish it from the typhoid fever, so attentively studied in our day. The typhus of the Crimea has solved the question of identity or of non-identity between the two diseases, and it is no longer possible to confound them, although they have more than one bond of relationship, and an apparently common cause.* It is generally agreed that typhus originates in an animal miasmatic intoxication, resulting as well from too great a number of human

---

* See the Memoir published June 2, 1856, in the Comptes Rendus of the Academy of Sciences. The observations which I collected in that Memoir have been since reproduced by writers who have forgotten to mention where they found them. I do not complain, since they have aided me in spreading the truth.

beings closely shut up, as from the putrid decomposition of animal matter. Consequently this disease appears in vessels, barracks, camps, prisons, and field hospitals, crowded with the wounded, whose sores are a source of abundant suppuration. It shows itself in besieged cities, and in certain localities infected by dead animals, or men left without burial. There is this distinction between the two diseases: Want is the essential cause of typhus, while it is scarcely an accidental cause of typhoid fever. Authors agree on the non-relapse of typhoid fever. Two physicians, Messrs. Lardy and Laval, were attacked with typhus, although they had four or five years before had the typhoid fever. There were found traces of the latter in the cicatrices of intestinal ulcers in M. Lardy, who, less fortunate than M. Laval, died. This is another proof of the non-identity of typhus and of typhoid fever.

Contagion, still very much doubted in the latter, is not in the former. We have seen, especially in the service of M. Lallemant, major physician, the typhus propagated from bed to bed in the wards, spreading itself by contact, and killing those sick of some light affection. At another time, as in the field hospital of the first division of the third corps, the typhus attacked almost every hospital employé; fifteen physicians in sixteen were attacked, and there did not remain a single healthy attendant. The word contagion, as applied to typhus, ought to be explained. The typhus, generated spontaneously by certain definite causes, does not transmit itself by the contact of a sick person with one in health, but rather by infection; that is to say, by the air charged with the elements of typhus. The morbific miasm exhaled from the surface of the sick, or from animal detritus, infects the man who breathes it; and once absorbed, during a longer or shorter time, called the period of incubation, it prepares the organism for sickness.

Typhus differs on one point from most epidemic diseases, such as small-pox, scarlatina, measles, sweating sickness, cholera, etc. These are brought on by a predisposing condition of the atmosphere, and the physician has no means of preventing their invasion. On the contrary, the causes of typhus are known, to such a degree, that we can produce or terminate at will the typhic influence. Another remarkable difference between typhus and common epidemic maladies, is that the latter have a regular course of duration, while typhus holds on and extends its ravages indefinitely, so that by any well directed measures we cannot master it.

Typhus breaks out more or less quickly, according to the intensity of the affection, and the resistance of the organism. Every sick person sends out noxious emanations. When the wards are full, when the cases of primary or contracted typhus increase, the epidemic furnace acquires a greater energy, and its manifestations radiate through the entire personnel of the hospital. It was in this way that the sisters, the chaplains, the physicians, and the attendants were so cruelly struck down during the war of the East. We have seen some physicians, less predisposed, endowed with a greater power of reaction, or of elimination of the absorbed miasm, overcome the epidemic influence in a manner little apparent, but still real. Each time that the fire of infection increased in the hospital by swelling the numbers of the typhic, they were seized with headache and sleeplessness, the tongue was dry, and the countenance assumed a typhic aspect. These symptoms held on three or four days, when the typhic veil was torn, and they returned to their usual health. Sometimes, however, the morbid state continued, and then the issue was almost always fatal.

The progress of the typhus in the Crimea was less uniform and regular than that so well described by Hildenbrand, one of the most eminent physicians of the Vienna school.* The irregularity of typhus in the Crimea was due to various complications, chiefly of scurvy, dysentery, and intermittent fevers. It was on the 1st of January, 1856, that the typhus, which the year previous had commenced to show itself, was greatly developed.

* In a treatise upon contagious typhus, published in Vienna in 1810, and translated into French the next year by M. Gasc, health inspector in the army.

During the last days of the siege of Sebastopol, the hospital gangrene, that typhus of wounds, committed great ravages. The scurvy, already remarked by Franck as the precursor of typhus, had assumed immense proportions. The contagious typhus waited only for the concentration and accumulation consequent upon the rigors of winter, to break forth. The soldiers packed into their tents, hermetically closed, in which the ground was damp and impregnated with impurities, were fatally poisoned by the organic miasm. Moreover the energetic excitement, so powerful in resisting typhus, fell with Sebastopol, and they saw themselves delivered over to the epidemic without relief from these powerful moral reactions.

The typhus described by Hildenbrand should have shown the regular character assigned by that writer, if not among our soldiers, broken down and already a prey to other maladies, at least among the physicians, the chaplains, and the other attendants of the hospitals at Constantinople, whose constitutions were not broken; but even here irregularity was the rule, and the eight stages described by Hildenbrand were not perhaps observed in a single case. The premonitory state (lassitude, unrefreshing sleep, lumbar pains, shudderings, painful tension in the head, and vertigo), so common in typhoid fever, were often wanting. The typhus almost always began with a chill and an inflammatory period, which foreshadowed—besides the more or less distinct catarrhal state of the eyes, nasal passages, and bronchia—an intense frontal headache, dizziness as in intoxication, stupor, great prostration of strength, intense thirst, often a saburral condition of the digestive passages, and a calm or furious delirium. The skin became burning, and after two or three days was covered with a kind of eruption which was never absent except in subjects too much broken down, and which differed essentially from that of typhoid fever. This eruption appeared upon the trunk and limbs, in irregular groups of rounded spots, of deep red, not in relief, smaller in size than a bean, not disappearing upon pressure, and not of such a character as to be confounded with the spots of typhoid fever.

The continuity of the fever with 100 to 130 pulsations to the minute, was often broken by one, or more rarely two regular paroxysms, in twenty-four hours, which much resembled the accession of remittent fever, and gave to the typhus of the Crimea a peculiar character. The belly was soft, without pain, or flatulence, or the rumbling in the right iliac fossa, which is the proper symptom of typhoid fever. Constipation has almost always occurred, in place of the intestinal flux of typhoid fever, when dysentery did not exist upon the invasion of typhus. After an inflammatory period lasting five or six days, there supervened a nervous period marked by ataxic or adynamic symptoms, and often by a mixture of both. The nervous period lasted four or five days, and was but little observed when convalescence began.

The typhus sometimes ran through these three periods with a fearful rapidity, and death often supervened the third day, and even the second or the first. The typhus was then terrible. It rarely lasted beyond fifteen days, unless complicated with organic congestions of one of the three splanchnic cavities (the brain, the chest, and the abdomen), and a return to health almost always occurred, if at all, within twelve days. The sick passed suddenly from death to life, and the typhic veil of the face fell off, the eye became clear and intelligent, the appetite ravenous, and strength returned with great rapidity. Still the intellect retained the stigma of typhus, as shown by disturbing dreams by night, or delirium upon some points by day, although the reason was clear on the others. A weakness of hearing and sight, a more or less complete loss of memory remained for a considerable time, nevertheless the hair did not fall out as in typhoid fever. These favorable changes were often preceded by bleeding at the nose, sweats, and morbid condition of the urine, and sometimes inflammation of the parotid glands. It will be noticed that convalescence, which is so slow and so difficult to manage in typhoid fever, proceeds rapidly in typhus. The departures from a strict diet are little to be feared, which is explained by the absence of the lesions of the intestinal follicles, and of the engorgement of the mesenteric glands which form so constant and important a feature in typhoid fever, and which hundreds of post-mortem examinations have never discovered in our hospitals of the East.

To cure typhus, it is necessary, above all, to have pure air constantly renewed; to withdraw the sick from the causes of the infection, to ventilate the room, and order frequent aromatic and chlorine fumigations; to respect the inflammatory period as the greatest effort of nature; to drive out the miasmatic poison by an exanthematous pressure. Not to bleed unless the subject is very strong; if threatened with cerebral apoplexy, prefer generally leeches behind the ears, or cups between the shoulders to a general bleeding, a remedy of which we should be extremely cautious. Adopt the same means when the smallness of the pulse denotes an oppression of the vital forces, which are relieved by moderate sanguineous depletion. When from the beginning, as in the typhus of the Crimea, there were remittent paroxysms, it was well to attack the disease with a few doses of quinine. By this means, the continuity of the fever was secured, and it fell of itself after a few days, unless maintained by an organic congestion produced by its first attack. This complication frequently occurred when we did not take care to annihilate entirely the paroxysms, that is, the recurrences of the fever. On an attack of typhus an emetocathartic was useful, especially if there existed some gastro-intestinal difficulty; we give mucilaginous or acidulated drinks, and sometimes wine in water. In the nervous period we have recourse to the remedies employed in ataxy and adynamia. In the latter, tonics, such as Malaga or port wine, much hastened recovery.

Such is a brief statement of the treatment which gave the best results in the army of the East, and upon which the most experienced physicians relied, such as M. Cazalas, who from the first conceived the idea of regulating the inflammatory action by quinine, by expelling the paludal element, which had so marked an influence upon the sickness of the Crimea. To sum up;

the typhus reveals its true nature by its infectious character, its easy communication, the rapidity of its progress, the whole train of its symptoms, and the absence of anatomical lesions.

We may derive still further light concerning the typhic affection, by comparing the typhus of the Crimea with epidemics of the same kind that have afflicted populations and armies, at other periods. Doubtless there is no absolute resemblance, because, as we know, epidemic manifestations vary in the same disease, according to the season, the locality, and the people, but we find in the typhus of the Crimea, the same putridity and rapid destruction of the vital forces, that distinguish the typhus of Mayence, the delirium, stupor, and rosy exanthem described by Hildenbrand, etc. If the typhus of the Crimea was not very grave, as compared with the dreadful epidemics of Mayence and Torgau, we may explain this by the different conditions in which we found our army; a better hygiene, the prompt care bestowed upon the sick, the facility and abundance of means of transportation, the multiplication of hospitals, and lastly to the moral conditions and material resources which did not exist in the campaigns of 1812 and 1814.

The appearance of contagious typhus was the most terrible event that happened to the army of the East. The blow struck first in the Crimea. In November, 1855, the reports made to me by M. Scrive, noticed in that month 11 cases of typhus, of which 6 were fatal. December showed the following increase: from the 1st to the 10th, 4 new cases, 1 death; from the 10th to the 20th, 46 new cases, 21 deaths; and from the 20th to the 31st, 92 new cases, 41 deaths. During the month, the field hospitals of the Crimea had emptied upon the hospitals of Constantinople 3,206 sick. All of these had, in the field hospitals of the Crimea, been in contact with the typhus, had lain in the same wards side by side. Some were even lightly tainted with typhus, so that the hospitals of Constantinople began to show from the 10th to the 20th of December, 13 cases and 2 deaths, and from the 20th to the 31st, 28 new cases and 7 deaths.

From that moment, I sought to restrain from the start, by an extensive plan, the first menaces of this new disease, and submitted to Marshal Pelissier a project having for its end the maintenance until spring, in good quarters in the camps of Constantinople, of all the sick who came from the hospitals of the Crimea; and likewise of sending thither several thousand men from the regiments, who were drooping, or already in the regimental infirmaries.

"I, myself, consider the first project," the Marshal replied on the 4th of January, "as one of easy application, and as capable of giving good results by allowing the complete recovery of men still weak and subject to relapse. Yet I do not think we ought to adopt it unless with men whom the health officers recognise as out of condition of rendering service, and whom they judge necessary to send to Constantinople for the winter. When these shall have gained sufficient strength to return to the ranks, they should rejoin their corps, as heretofore. I have therefore the honor of informing you, that I have issued orders to General Larchey, to put this measure into execution, in this qualified sense. I have likewise examined the proposition which formed the subject of your second note, dated on the 1st inst., relative to our sending to Constantinople the drooping men now in the Crimea. As to these, its adoption appears to me less opportune, on account of the great numbers already sent off. Those thus sent away, are already a too active cause of the diminution of our effectives, and they realize indirectly the end that you propose. I, however, am perfectly conscious that a solicitude for the well-being of our soldiers alone prompts you to make the proposition, and I thank you for having done so."

The 4th January, 1856, M. Scrive informed me that the health of the Crimean troops continued comparatively satisfactory. "Your prescriptions, sir," he said, "are being executed, and you shall have, as soon as possible, the information concerning the ambulances, which you require to be supplied from the heads of the department."

"From information received from the intendant-general, I learn that we have now in service 5,000 complete beds in barracks, and 5,000 places for men in tents. There are no troops in shelter tents; all have either camp tents or huts. The Zouaves' huts answer their purpose well; the rain enters very little, and the ingenious soldier immediately places a sod over the holes. The large heating apparatus for Aurelle's and Herbillon's divisions do not answer; they have been inundated by the deluging rains. The partially underground tents are healthy; those of the flying ambulances at the Tchernaya are perfect models. The regimental infirmary barracks are being constructed, but not without great difficulty. I hope, however, to have two ready for each regiment in a few days. In fact, we shall soon be well provided in every way."

"It snows and rains continually. The sun has not shown itself for a fortnight. In the face of such a severe winter, the health of the troops is better than could have been anticipated."

A few days later, M. Scrive, in a report addressed to the Council General of the Health of the Army, said:— "The causes which have produced this healthy condition of the army were foreseen, and the result was certain. Several months ago, Inspector Baudens wrote to the Minister of War, that the hygienic condition in which the army would be placed during the winter would render sickness more general and severe. This prophecy was unfortunately substantiated. Although the troops have not had to experience the fatigues of a siege, yet they suffered from the condition of continued warfare which followed the fall of Sebastopol. It was no longer a circumscribed circle of trenches, but a line of defences, thirty-six miles in length, from the valley of Baidar to Sebastopol, which had to be defended. Continual alarms obliged three-fourths of the troops to remain under arms. This extended line of defence rendered numerous pickets and advance-guards necessary; it made the furnishing of provisions difficult and irksome; and it was spread over ground in many instances unfavorable for winter camping, but rendered necessary by war. , , , ."

The hygienic conditions of our hospitals at Constantinople and the appearance of the new epidemic, however, actively engaged the attention of our head physicians. I had frequent conferences with them, and visited them almost daily during their rounds, so that I might perfectly understand the situation. They also represented to me, in writing, the cases where an ameliorated condition was urgently required.

The suggestions I had made to the Commander-in-Chief brought me, on the 9th January, 1856, the following despatch from M. Angot, military superintendant at Constantinople, relative to the establishment of a depôt for convalescents:—

"Sir—The General of Division commanding, orders me in a letter, dated 8th January, to confer with you, in accordance with the orders of the Marshal commanding, respecting the immediate establishment at Constantinople of a sort of depôt for convalescents, destined for the use of the soldiers now in the Crimea who are too weak to withstand the severities of the winter.

"This measure being adopted on your suggestion, I have the honor to request you to inform me what conditions of location you deem most applicable for the immediate establishment of this depôt, as much with regard to the medical attendance as for the adoption of particular regulations, if such are required.

"It is indispensable for me to know your opinion on these questions, before indicating to General Larchey the system to be followed in realizing to the utmost the intentions of the Marshal commanding. You will, perhaps, with me, recognise the advantage of a conference, in which Dr. Thomas and the military sub-intendants shall be asked to take part. In this case, I shall be obliged by your informing me the day and hour which would be most convenient."

The 11th of January, a committee, composed of General Larchey (President), of General Pariset, of Military Intendant Angot, and inspector Baudens; De Missy, sub-intendant; Lieut.-Col. Cadart, Chief of Engineers; M. Thomas, Physician-in-Chief; and having for its Secretary, Lieut. Col. Gravillon, Major of the Staff, was assembled in conference, and made many arrangements of recognised utility. These formed the basis of detailed instructions, to be found in the appendix.

Notwithstanding the precautions taken, the typhus made, both in the Crimea and at Constantinople, rapid and alarming progress. In the Crimea, there were counted from the 1st to the 10th of January, 34 new cases, and 49 deaths; from the 10th to the 20th, 164 cases, and 70 deaths. In the same period at Constantinople, there occurred, from the 1st to the 10th, 95 new cases, and 15 deaths; from the 10th to the 20th, 162 cases, and 32 deaths; and from the 20th to the 31st, 205 cases, and 18 deaths.

The epidemic was sadly confirmed. There can be no doubt as to the cause which propagated it in our hospitals at Constantinople. It was the impossibility of isolating the men tainted with typhus, and the crowding.* In a report made to me January 29, 1856, by Doctor Garreau, the Physician-in-Chief of the hospital of Daoud-Pacha, he thus relates how the accumulation of the sick in that hospital kindled suddenly the typhus:

"At Daoud-Pacha we enjoyed very satisfactory sanitary conditions up to December 21st, but diseases of bad type were not slow in appearing, in proportion as the sick sent from the Crimea became more numerous, especially when this number passed certain limits. A care-

* Physicians and Directors do not agree upon the word crowding. The latter apply the definition literally; so that, in a hospital prepared, for instance, for 1500 sick, they would not, in reaching that number, especially if each patient had 70 cubic feet of air to breathe, be crowded. But with the physician, a crowding exists, when an increasing mortality reveals the presence of contamination from a collection of sick. From that moment he should advise a reduction in number of patients, and the disinfection of the wards. In a campaign, when a soldier is convalescent, he leaves his place for another more sick than himself, and the beds are never empty day nor night. As each sick man is a furnace of mephitic emanations, we may see how an encumbrance may be rapidly gained. In times of peace, a hospital for 1500 sick has scarcely 1000 beds in use at any one time, and a third of the convalescents walk out into the courts and gardens during the day, thus benefiting the remaining sick, by giving them the 70 cubic feet of air allowed them in the wards.

ful observation of facts demonstrates satisfactorily, that crowding, in badly ventilated and poorly lighted wards, without fires, was the cause, I do not say the only but the principal cause, of our actual condition.

"On the 21st of December, we reckoned 525 sick, all doing well. We had 729 beds occupied on the 1st of January, and the sanitary condition was passable, because the defective wards were not as yet crowded. From the 1st to the 10th of January, our number increased to 877, and from the 10th to the 20th it amounted to 1,140 scorbutic or diarrhœa patients, almost all of them quite debilitated. It was at exactly this time, when it was necessary to crowd the beds together, and exceed the normal number, that the typhus fever began to rage in the hospital, beginning, as you know, in the upper story. Within fifteen days more than 125 cases of continued fever were disclosed, the persons attacked being almost all of them convalescents. About eight in ten came from the Crimea. The sick were few on the 11th, more numerous on the 21st of January, and increased rapidly from the 21st to the 29th; following step by step the progress of the crowding in the wards, for the sick placed in the vast galleries, where the air was constantly renewed, did well."

The instructions which the Minister of War had given me in writing had anticipated the terrible and exceptional periods. He said:

"Independently of your duty as inspector, which invests you with the right of direct correspondence with me, I desire that the authority of your rank as well as your surgical skill, may be employed for the benefit of the army, and I hope that you will not remain unconscious of the impetus which might be given to the medical service by your presence during your sojourn either at Constantinople or in the Crimea.

"To this end, when you think it expedient, or when the circumstances may require it, you are to take temporarily the direction of the service, and to practise as well on the battle-field, as in the field and the regular hospitals, and both by your example and your counsels to support our military physicians in the heavy task resting upon them, and sustain them in the way of zeal and devotion for the cause in which they are engaged.

"If any grave diseases should break out at one or many points in the country occupied by our army, you will search for the causes, and study their character, symptoms, and progress, suggest the preventive measures which circumstances seem to you to require, and give to the health-officers such instructions for their guidance as may be proper in the treatment of these diseases.

"You will give to the general commanding-in-chief, and to the intendant general as well as to General Larchey, and the intendant of the station at Constantinople, all the information which they may ask, and such advice as you may think fit to submit to them, relative to the health of the army, upon the salubrity of grounds and localities, upon the establishment and organization of hospitals, field-hospitals, etc.

"For the rest, my dear sir, I leave it to your solicitude for the welfare of our soldiers, and to your experience in campaign service, to supply what the foregoing instructions have left incomplete, and I feel assured that you will find in the command and administration, that cordial concurrence, which will render your mission as fruitful of results as possible."

During the whole course of the epidemic I took the official direction of the health service in the army, and was thus able to impart to it greater unity and strength. I entered then upon my duties of Inspector, which placed me in a more elevated sphere, as the delegate of the minister.

It was necessary to employ energetic measures, without which the mortality would have been unlimited. The principal remedies were the isolation and ventilation of the sick. I earnestly pressed the military intendant to place the typhic patients in special wards, where the air might be freely admitted, and thus at the same time guard the other sick against the risk of infection. It became necessary to open new hospitals in temporary barracks, to prevent encumbrance, to find places for 5,000 sick, and to be able to lodge four typhic patients only in each barrack in the camps of Maslak, instead of eight, as in common sickness. Our English allies tendered us every kind of assistance, both personal and material.

General Storks offered to go, and prepare in one of our camps a hospital complete for 1000 sick, and to furnish and attend them, if we desired it. "Whatever we do," said he, "we can never repay the French, for what they did for us last year." Happily we were plentifully supplied with materials, and the military intendant at once introduced a salutary change in the rations. The things most needed, were room and pure air, and I urged the adoption of temporary barracks. This was made the subject of several consultations under the presidency of General Larchey, and it was decided to separate the sick, and to increase the number and extent of the hospitals.

At the same time I was occupied with devising measures for preventing the spread of the pestilence in the Crimea, and on the 11th of February, 1856, I wrote to Marshal Pelissier and to the Minister of War as follows:

"Reports from the Crimea show a great increase in the numbers of soldiers entering the hospitals, while at the same time diseases are reducing our number of physicians. I pray your Excellency to permit me to submit, on this occasion, some reflections inspired by an interest for the service.

"An increase of the sick was foreseen. An army whose effective strength is doubled within a year, and which includes a great number of young soldiers, would, as I have said in former reports, send during the rigors of winter large numbers to the field hospitals. In view of this probable event, each regimental infirmary received two barracks, and 6,000 mattresses have been sent to the Crimea to replace the mats upon which the sick have lain. These prudent measures have nevertheless not allowed us to completely dispense with putting the sick in tents, where they have no other bed than a mat, and one or two blankets; many have had their feet frozen, and have reached us from the Crimea in the most pitiable condition.

"I presume that recourse was only had to the tents in

cases of absolute necessity, and it is well to remember that this shelter, so excellent in summer, may become very pernicious in winter, when they are hermetically closed to exclude the cold, and the air not being renewed, becomes loaded with noxious miasms, that develop typhoid fever, if not the typhus. When these are opened, the frost seizes upon the anæmic men within.*

"The fault of arrangement of these tents has been their nearness to each other. They should not occupy more than a third of the ground, so that they may be moved daily when the weather permits, and keep at a distance from each other the hotbeds of infection of which they so soon become the seat.

"The field-hospital barracks, encumbered with soldiers dangerously sick, are rapidly infected. They are unfortunately much too near together, so that they mutually radiate upon each other their infection. To render

---

* Living in common in tents, carefully closed, is extremely pernicious to the health.

In this confined atmosphere, vitiated by the vapors thrown off from the breath and skin, and by the emanations of the infected soil, the soldier yields to the effects of poison by putrid intoxication. We know that confined air may contain, besides its constituent elements (oxygen and nitrogen), traces of the oxyde of carbon, sulphuretted hydrogen and ammonia, and that it is besides altered by an excess of carbonic acid, and other principles called miasms, not as yet appreciable by chemical reagents, but whose presence is revealed by their disastrous effects upon human health. From the experiments of the savan M. Dumas, it appears, that an adult changes into carbonic acid gas, by respiration, in one hour, all the oxygen in 3·2 cubic feet, and that to supply this loss, twelve cubic feet of fresh air are required per hour by each individual.

Carbonic acid gas being less respirable and heavier than the elements of air, we learn the danger in soldiers passing the night together, under an air-tight tent, and lying upon the ground. If asphyxia does not occur, as in the Grotto del Cane, near Naples, the aerization of the blood is impaired, and it becomes impoverished. As to the mephitism caused by organic emanations, it carries into the circulation a special poison of more or less active character, of which the prolonged effects appears as typhus in campaigns, and in times of peace, when the air is not well renewed in the hospitals these influences result in other diseases, such as typhoid fever, scrofula, and consumption. According to some hygienists, the indispensable ration of pure air to each man is at least 18 cubic feet per hour. (*See Appendix.*)

them healthy, they should, besides the measures already advised, be fumigated five times daily; twice with chlorine, at six o'clock in the morning and at seven in the evening; and three times with dry sage, after the custom of the Turks, who in their hospitals take effectual precautions against contagious diseases. The aromatic fumigations should be made, one at noon, and the two others half an hour after the fumigations with chlorine. These precautions had been entirely neglected, even in our hospitals at Constantinople, and I was obliged to enforce them with vigor, but not without difficulty. The regimental infirmaries, and if possible the tents of the soldiers, where the trampled ground will not allow of a change of place, should also be purified by these fumigations. While the liberated chlorine decomposes and neutralizes the miasms, air charged with aromatic vapors, searches into the corners and recesses, and escapes, bearing with it the sickly odors. They are, so to speak, an atmospheric broom.

"Dried aromatic plants abound in the markets of Constantinople, and a supply might easily be sent from there to the Crimea.

"Your Excellency is aware that the typhoid fever, and especially the typhus, makes some progress before Sebastopol. Thanks to the measures in operation, the latter does not threaten us with the fierce ravages which it displayed in 1812 and 1814, but nothing should be omitted in watching and preventing the spread of this powerful epidemic.

"The accumulation of the sick from the Crimea, in our hospitals at Constantinople, operated unfavorably upon their hygienic condition, in consequence of the severity of the diseases. Commonly, among a hundred sick, only ten are in danger; but here the proportions are reversed, and in a hundred tainted men, ninety are in a perilous condition. There are scorbutic patients, with infected breath, and dysenteric patients with contagious typhic emanations. The miasmatic intoxication of our hospitals has developed numerous unlooked-for troubles. To arrest them, I have proposed the reopening of the discontinued hospitals, and the creation of others, sufficient for about five thousand beds, in the camps at Maslak, and between the hospitals of Ramis-Tchifflick and Maltépé, where the sick may be free, and in a salubrious and well-ventilated situation.

"To prove that to check typhus it is only necessary to place the sick in a non-infected place, we may refer to the depot of convalescents prepared in a portion of the camps of Maslak. . In one thousand men, not one showed any traces of typhus; they recovered rapidly, and only fifteen soldiers were sent back to the hospitals, and nineteen to the infirmary.

"Mytilene would have been a providential resource for the scorbutic, on account of its fine climate, thermal waters, and abundance of fresh vegetables.

"Your Excellency knows that at Constantinople there are barracks for lodging about twenty-five thousand men, that might be easily converted in twenty-four hours, into good hospitals. These resources would allow of the arrangement providing for the sick sent from the army in the Crimea, according to the views that I proposed. The ailing fill the regimental infirmaries; but it would be better to send them directly to the quarters prepared for their reception, without their passing through the field hospitals, where their condition is aggravated.

"The medical employees in the places devoted to contagious diseases have paid a large tribute to the typhus; many physicians are dead, and twenty-five, not including those actually sick in the Crimea, are under treatment in the hospital at the Russian embassy.

"Your Excellency will observe with satisfaction, that not a single army officer has shown any traces of the typhus, which proves that it is, as it were, imprisoned in the hospitals, and is not propagated in the camps by contagion, although it is generated spontaneously in the tents."

M. Larchey, General of Division and Superior Commandant at Constantinople, to whom I had imparted my plan of converting into hospitals the camps of which I have spoken, directed me to prepare some instructions to lay before a commission, which met February 13,

1856. The minutes which follow repeat very nearly what I have already stated.

Upon the call of the General of Division, the military commandant—Messrs. Pariset, brigadier-general, attached to the general of division; Angot, military intendant, upon special service; Baudens, inspector of the health service of the army of the East; De Missy, military sub-intendant of the 1st class; Cadart, lieutenant-colonel commandant of engineers; Wilson, military sub-intendant of the 2d class; Thomas, principal physician of the 1st class; Demortain, principal apothecary of the 2d class—assembled in council February 13, 1856, at the general head-quarters, for the purpose of inquiring into the measures to be taken to prevent the development and spread of typhic diseases in the military hospitals of Constantinople.

After full discussion and inquiry at this session, the following arrangements were adopted, namely:—

1st. The numbers of sick in the hospitals are to be reduced, until the miasmatic infection shall have disappeared.

2d. An examination is to be made by the commandant of engineers relative to the labors necessary to convert into hospitals the barracks not occupied by troops.

3d. The vigorous measures recommended in the order of April 13, 1855, and reiterated in the order of February 11, 1856, relative to the ventilation and fumigation of wards, are to be put into effect.

4th. In addition to the chlorine fumigations, which should be made morning and evening, as prescribed in the formulary of medical hospitals, it is recommended to make three others daily, with aromatic plants, such as dried sage, after the manner of the Turks. These aromatic fumigations should be made morning and evening, a little while after the chlorine fumigations, and also at noon.

5th. Typhoid fever, typhus, and dysentery, being furnaces of infectious emanation, it is necessary to fumigate with chlorine all the clothing and bedding of persons dying from these diseases, and also to subject to fumigation (if the sick can change their beds) all the bed clothing used during the course of the disease.

6th. When the weather is so mild that the patients can go out, it is ordered to fumigate the empty barrack throughout with extra quantities of chlorine, taking care to hang out the mattresses and blankets; to keep the ward empty as long as possible, that it may be purified and cleansed: and to do this in rotation through every ward of the hospitals.

7th. The doors and windows are to be opened when the weather will permit.

8th. Redoubled care should be taken with regard to cleanliness—1st, of the sick, by washing their feet when they are brought in, rubbing with warm water and soap their thighs, legs, and arms, when it is not possible to give them a bath, or when the bath should not be used; but to do nothing without the previous advice of a physician; and 2d, in the wards, corridors, and privies. Disinfectants are to be placed permanently in places for evacuations, and a pinch of sulphate of iron is always to be left in the night vessels, after being thoroughly cleansed.

9th. A wash-house should be built at Gulhané, and another at Pera; the former to be devoted to washing the linen of the hospital at Gulhané and at the University, the other that of the hospital at Pera and the Parade ground.

10th. Smoking or spitting in the wards to be prohibited. There shall be assigned a place for smoking in each hospital, and wooden spit-boxes shall be placed at regular distances. The attendants of the infirmaries must not spill any drinks, broth, or other liquids.

11th. On account of the increase in the number of the sick, and the diminution in the numbers of employees in consequence of death or sickness, and in view of the good results already attained from the employment of infirmary-dressers, there shall be designated to each hospital a certain number of intelligent assistant attendants, to keep the visiting pass-books and perform the minor dressings.

12th. A commission, consisting of General Pariset, of

the sub-intendant charged with the hospital service, and Lieutenant-Colonel Cadart, commander of engineers, shall visit in succession all the hospitals at Constantinople, and propose such improvements as circumstances may require, and which are not above indicated.

During the month of February, the number of sick in the Crimea amounted to 19,648, of whom 2,400 died, 1,993 recovered, and 8,738 were sent to Constantinople; and during the same month, in the hospitals of Constantinople, the sick amounted to 20,088, of whom 2,527 died, 3,617 recovered, 649 were sent to Gallipolis and Nagara, and 3,717 to France.

The plague in Egypt, in 1792, is spoken of with terror. Said the illustrious Desgenettes, in his *Medical History of the Army of the East*, " from the most accurate and careful reports, the army in Syria lost by this epidemic about seven hundred men." Our typhus committed much more dreadful ravages.

Notwithstanding my urgent representations, a sufficient number of places for the increasing numbers of typhic patients were not provided; and on the 28th of February I addressed the following letter to the Minister of War :—

"To HIS EXCELLENCY THE MARSHAL.—The progress of typhus continues upon the increase, and 150 new cases have occurred in the hospitals at Constantinople within twenty-four hours. At Maslak No. 1, among 420 sick, 180 have the typhus, and at Ramis-Tchifflick, among 700 sick, there are 250 cases of this disease.

"Your Excellency will observe that there exists in certain hospitals a serious tendency to the disease, and that it demands a prompt remedy. The remedy is simple ;—air, pure air, constantly renewed; and to secure this, we should, as quickly as possible, remove one half of our hospital population to the unoccupied barracks of Maslak, and there construct a great encampment and bivouac. This has been my request, morning and evening, daily.

"On the 1st of March they promise me 2,000 places in barracks; but these will be inadequate; the more so, on account of the great number arriving from the Crimea.

Yesterday, 650 sick arrived from Eupatoria. I have visited them, and they are properly placed on litters and furnished with blankets. The passage only occupied thirty-six hours; four died at sea, and only ten men, one of whom was of the medical staff, had the typhus.

"An error has spread among the authorities, which I have endeavored to counteract, because it might be attended with pernicious consequences. This is, the comparison of typhus with cholera, and the belief that the disease will disappear of its own accord. The cholera, whose cause is unknown, sweeps forward with a force that nothing can check, arrives at its maximum, and then declines and spontaneously disappears. Typhus, on the contrary, which we know is brought on by privations, continues until the cause is removed. Its element is human miasm, rendered contagious and virulent, in proportion to the number of typhic patients assembled in one spot.

"It was because the great measures of prevention were neglected, or could not be employed, that typhus, during the wars of the empire, was seen to take such an extension, and strike terror into the hearts of cities. So it was during the siege of Mayence. The wise foresight of the government, and the great solicitude of your Excellency, has happily not left us disarmed in face of the epidemic. We have barracks for 25,000 soldiers awaiting inmates. Let us hasten to use them. But opening barracks to satisfy new wants, on the gradual arrival of our sick from the Crimea, will not meet the end required; it is simply waiting till the approaching tide swallows us up. And yet I should say, that the brave and worthy General Larchey has the best disposition to aid in any measures, as also has the Intendant Angot, and I have only to approve my excellent relations with them. Why don't we act more rapidly? There are, apparently, difficulties in the execution, of which I am not cognisant. Thus, I have heard the Intendant object, in one of our general conferences, against my plan of substituting field hospitals for the large general hospitals, that the Minister would not allow the establishment of them out of the Crimea. I admit that the instructions are plain, to one who has no responsibility; so I dare not complain, and can only deplore the situation in which I am placed.

"I have already informed your Excellency that General Larchey has permitted us to select, from among the soldiers about to be sent to France on sick leave, two hundred auxiliaries for the hospital service, to whom we will entrust the visiting pass books and the charge of the minor dressings. This measure assures the perfect fulfilment of the service; and as soon as the auxiliaries recover their health, we will send them back to the Crimea, and supply their places with others. The following are the last daily reports of the progress of the typhus in our hospitals at Constantinople:—

	Remaining.	Recovered.	Died.	New Cases within 24 hours.
February 25,	1,706	15	40	166
" 26,	1,826	7	46	188
" 27,	1,846	50	56	137

"Three military physicians fell victims to the typhus on the 26th; and it pains me to inform you that one of them, M. Sagne, Aide-Major Physician, in a moment of delirium, inflicted upon himself several cuts with a razor, which opened the arteries of both arms, and he died of the hemorrhage that followed.

"M. Gérard, Major-Physician, died at the hospital, at the Russian embassy; and the principal physician, Volage, has been poisoned by the breath of the sick, upon whom he lavished the treasures of his science and self-denial. His death has caused the keenest regrets.

"Within three days six new cases of typhus have appeared in the medical corps at Constantinople, and yet no one hesitates, and each will discharge his duty till the end.

"P.S. General Larchey, to whom I have communicated this report, has brought to my knowledge a ministerial dispatch, this moment received, in which your Excellency has ordered the ailing soldiers of the Crimean regiments to be sent to Constantinople. This measure would have been well when I advised it; but the ailing are to-day the sick. We are going to open wide to them, the doors of the camps of Maslak, which soon, I trust, will be converted into hospital establishments."

The courier, on the morrow, bore another dispatch, which I addressed to his Excellency the Marshal Pelissier.

"CONSTANTINOPLE, *February* 29, 1856.

"To HIS EXCELLENCY THE MARSHAL.—I have the honor to address to your Excellency a report upon the progress of the typhus in our hospitals at Constantinople, which will enable you, with the documents transmitted directly from the Crimea, to understand the sanitary condition of the army. I cause a report to be made to me daily of the number of men tainted with the typhus, of the new cases appearing during the last twenty-four hours, the number of recoveries, and the mortality.

"These reports are instructive, and the following table gives their returns since the 20th of the present month:—

February	Number of Typhus patients.	New Cases within 24 hours.	Recovered.	Died.
20	1,450	62	4	36
21	1,517	111	10	44
22	1,586	168	9	62
23	1,556	105	27	41
24	1,615	128	8	43
25	1,706	166	15	40
26	1,826	188	7	46
27	1,848	137	50	56
28	1,927	168	39	50
29	1,969	178	79	60

"The epidemic wave is rising, and we can only escape it by watching its conditions, and raising against it an effectual barrier. It is only necessary for you to cast your eyes over the tables showing the new cases occurring within twenty-four hours to convince yourself of the contagion of typhus, and of the necessity of adopting measures for limiting its spread as quickly as possible. It would be well to advise means to retain all the men tainted with typhus in the Crimea, if it could be done without exposing the army to too great risks. But in the mean time, it is easy to arrest the typhus at Constantinople, and to prevent it from spreading. We have providentially in our camps empty barracks for 25,000 men. They are arranged upon elevated grounds, and are in good hygienic condition. Let us remove thither half of our hospital population, some 5,000 sick, and I will answer for it that we will check the progress of the mortality from typhus almost immediately;—for what, sir, is necessary to cure this disease? Only an abundance of pure air, constantly renewed, in place of air charged with contagious miasms. Every form of medical treatment, whatever may be its virtues, will prove a striking failure, if the first essential condition, that of *disencumberment*, is not fulfilled. For this end, I ask only some field hospitals and beds. Straw beds, even, would suffice for probable wants, and the necessity for them would, I trust, disappear with the return of fine weather. This measure ought not to present great difficulties in execution. We shall have, in a short time, 5,000 places under barracks to satisfy the new wants as they shall arise; but in so doing, we allow ourselves to be pressed upon by necessity, we do not meet it. We shall find ourselves some day invaded by the sick, in place of having prevented the sickness.

"Pardon me, Marshal, for expressing myself to the point. I must tell your Excellency the truth. Myself a man of action, field hospital surgeon, I would wish to move off with some supplies, and with my sick to establish a great bivouac in the unoccupied camps. The administration, on which all responsibility rests, cannot, I am aware, act so quickly. They encounter, besides, great difficulties in recruiting auxiliary infirmary attendants.

" As for the *personnel* of the health service, although reduced by mortality and sickness, it can yet face probable events, if the commander, whose solicitude has never been dormant, still continues to second our efforts. We have here upon our hands 10,000 sick soldiers, who, in place of being sent to France on sick leave, would like nothing better than to become our infirmary attendants. They would otherwise be lost, so far as being of any service in the Crimea, during the next four months. By placing them in charge of the visiting passbooks and minor dressings, we may hold them in reserve, and as soon as they become fit to resume active service, we might send them away and replace them by others.

"I visited the sick on board the ships arriving from the Crimea, and presided over the arrangements for placing them upon steamers to be sent to France. I can affirm that every measure possible, suggested by humanity and affection, has been adopted. I have taken especial pains not to allow any sick to embark that were tainted with typhus; but it is not within the power of man to prevent the malady from breaking out upon the voyage. We have had for a year 2,500 places prepared for the sick in the lazaretto of Marseilles, 500 in that of Port-Vendres, and 1,500 in that of Toulon. Thus, in this way, everything has been foreseen by his Excellency the Minister of War.

"My relations with M. Scrive have brought to my knowledge the excellent arrangements adopted in the Crimea. It is of the highest importance that the site of the camps should be changed as soon as the weather will permit; but, in the mean time, it is necessary to empty and cleanse the field hospitals.

"Thus far the epidemic has happily been limited to the hospitals, without entering the camps. On the 26th, three physicians died of typhus, others are now in great danger, and twenty-four are under treatment. Everyone does his duty with self-denial and with redoubled efforts. We have already lost eight military physicians killed, and thirty wounded on this battle-field. I class among the *wounded* the physicians who have contracted the typhus among the sick whom they have attended. When the proper moment shall have arrived, I will with confidence appeal to your high benevolence concerning this class of officials, who since the beginning of the campaign have enjoyed the honorable privilege of conciliating more and more the earnest sympathies of the army."

Unfortunately, the weather became unfavorable. On

the 3d of March, 1856, the Intendant-general addressed me as follows from the general head-quarters:—

"I have read very carefully your report upon the sanitary condition at Constantinople. We are passing through a terrible crisis, and most of all need pleasant weather to enable us to pass through it in the Crimea; for a short continuance of fine weather would enable us to complete the purification of our ambulance hospitals. Already the old field hospitals of the 1st corps have been rearranged throughout, and are now perfectly healthy. We need to do the same thing everywhere; but the weather still continues so bad that we meet with the greatest obstacles. Meanwhile, we observe a marked improvement in the general health, and the number of daily entries has sensibly diminished during the last ten or twelve days. This may be due, in a great degree, to the oil, vinegar, potatoes, and mixed vegetables which I have caused to be distributed among the troops, so that there are but very few scorbutic patients now in the field hospitals. . . . . I think that the Marshal will give orders for the employment, to the greatest extent possible, of the barrack camp at Maslak for the use that you suggest. I am entirely of your opinion, that when possible, a seat of infectious disease should be abandoned without delay. I well remember that during the cholera at Varna I had to fight it in like manner; I required that the choleraic patients should be taken out of the hospitals, and placed in tents; I demanded pure air, and was answered that heat was needed. The number of sick increased, and they were at length forced to have recourse to the tents, and then learned that this was not so bad a plan after all. It appears to me that it is the same thing with typhus, except that large well-ventilated barracks, and only a few sick placed in them, would prove better than tents."

On the same day, a telegraphic dispatch from Marshal Pelissier, sent to Constantinople the orders necessary for the immediate establishment of field hospitals for 5,000 sick, and on that day I wrote as follows to the Minister of War:—

"The contagion continues its ravages. It will continue thus, so long as we cannot carry into the barracks of the unoccupied camps a third or a half of the sick in the hospitals. Of the 5,000 places that I have requested, 1,000 have been furnished, and we have thus been able to create a little space in our hospitals. This at once produced a diminution in the number of new cases. In fact, on the 1st of March, this number had fallen to 93; but, unhappily, this respite was enjoyed but for a moment, and new additions of sick, sent from the army before Sebastopol, have encumbered our hospitals, and have obliged us to encroach upon the rooms reserved for the sick who are dangerously ill. The number of new cases has thus been increased beyond any former period, amounting to 257 within twenty-four hours.

"By constant airing and ventilation of the wards; by five fumigations daily, two with chlorine, and three with aromatic herbs; by placing under the bed of each typhic patient a cup of chloride of lime; by scouring the floors and whitewashing the walls of the wards, one after another; by keeping always in the chamber vessels some sulphate of iron; by making large openings in the privies, for their thorough ventilation; by having, if possible, two beds for each man gravely affected by the typhus, and fumigating each bed for twenty-four hours after it is left; by washing the linen in boiling water; by an improvement in the regulations for the diet, and a more substantial broth, and Bordeaux wine for the sick—all of which measures we have regularly enforced—are we able to resist the pestilence, which nevertheless is gaining ground upon us daily. We shall be able to conquer it only when we shall have taken possession of the new hospital establishments prepared in the camps at Maslak. It is not without trouble that I can destroy in the mind of the command and the administration their feeling of security in the midst of so much danger:—they believe that the typhus brought from Sebastopol will disappear at Constantinople, when it ceases to be imported from the Crimea. It would result from this, that we need not trouble ourselves here on account of the epidemic.

In the meantime, the epidemic spreads rapidly among our hospitals at Constantinople. The only way to stop it is to remove half of the sick to empty barracks. Let this be done, and I will answer for it that the march and fatality of the typhus will be almost immediately checked. I ask only for field hospitals. The measure appears to present great difficulties in the execution. More places in barracks are promised, in proportion as new wants arise. By so doing, we wait till necessity compels us to act—we shall find ourselves some day invaded and powerless. I would wish to go with some supplies, and with my sick establish a great bivouac in the unoccupied camps."

At the close of a new report, written on the 6th of March, the minister addressed to General Larchey, at Constantinople, a telegraphic dispatch, in the following words:—

" The last report of M. Baudens makes me fear that the order for placing the sick in barracks, or under well-aired tents, has not been sufficiently carried out. Do everything that M. Baudens desires."

Night and day the health officers remained among their typhic patients, and scarcely left them, except to follow in the funeral procession and attend the burial of one of their number. Forty-six have perished, struck by typhus,[*] while bravely contesting its ravages, and eighty-two have died during the campaign. Never has the medical staff found a finer occasion for proving their traditional devotion to France, and to the army which has always regarded them with affection, and in the

---

[*] The names of the physicians and surgeons dying of the typhus were as follows:—*Principal Physicians*, Volage and Barby. *Majors*, Felix, Goutt, Bonnet-Masimbert, Frette-Damicourt, Moulinier, Braunwald, Girard, Rampont, Leclère, Pegat, Puel, Berthemot, and Peyrusset. *Aides-Majors*, Le Clère, Cordeau, Dulac, Savaëte, Gillin, Miltenberger, Perrin, Molinard, Gueury, Lasserie, Lemarque, Perry, Desblancs, and Masson. *Aides-Major Physicians*, Leker, Bouquerot, Servy, Demanet, Dartigaux, Ragu, Lardy, Forget, Sagne, and Fournier. *Sub-aide Physicians*, Jacob, Godquin, Sautier, and Labretaigne. *Aide-Major Apothecaries*, Boussard, Carron, and Granat.

ranks of which they have always been so eager to be classed.*

On the 2d of March, the population of Pera was saddened by the spectacle of three hearses, bearing in the same train the bodies of three physicians, who had fallen together, the victims of their own self-denial. These gloomy processions to the field of the dead break the spirit; for each one may truly ask himself: "Who of us to-morrow will receive this last sad adieu?" It fell to the share of the Medical Inspector to perform the painful duty of pronouncing the last words over the tomb of his unfortunate comrades.

The pious sisterhood of Saint-Vincent-de-Paul paid also a large tribute to disease, and thirty-one perished near their grateful sick, upon whose attendance they were unwearied; never showing signs of fatigue, nor disgust, nor anxiety about themselves in their delicate and assiduous attentions. Of those who died, twenty-four were struck by the typhus. The first one of them who took the disease was Sister Walbin, whose last words were: "The only favor I ask, is to be buried with the soldiers—they will be lonesome without me."

Meanwhile, instead of opening field hospitals, and ample quarters in barracks, they continued to send the sick to France. Within a month, 6,000 were sent thither in transports, and half the vessels, instead of returning to the Crimea, were despatched to Marseilles and Toulon. For the want of ships, the Crimea could not send us so many sick; and so the system continued as it was —the Crimea relieving itself upon us, and we upon France. Two hundred soldiers a day died between the Crimea and Constantinople; and the sailors, falling victims to the contagion, entered the hospitals along with the sick whom they brought. From Constantinople, the disease which infected the vessels was carried to Marseilles; and it would seem that we were indefinitely threatened with this real and terrible disaster. It was

* France knows how to appreciate every kind of heroism, and yet the widows of medical officers are deprived, by a decree that doubles the pensions of retreat for the officers of the army, of the advantages conceded to the widows of the latter.—AUTHOR'S NOTE.

a time for prompt action, under penalty of soon being rendered powerless—it implied the very existence of the army.

Two great measures were to be taken; the first was to send no more sick from Constantinople to France; and the second, to keep all the typhic patients in the Crimea, to the exclusion of the other sick, who might be sent to Constantinople.

I left for Sebastopol on the 9th of March, 1856, and at the moment of embarkation received a visit from M. Girette, the Director of the Imperial Mail Boat Company. He said: "The typhus is making such ravages in the ships of the company, infected by the constant transmission of the sick, that the mail service will be forced, in a few days, to be discontinued on the route from Sebastopol to Marseilles." Many of the sailors, firemen, and commanding officers of these vessels have died of the disease, and others are sick. M. Girette had not been able to replace these by others.

Immediately upon arriving in the Crimea, I visited a part of the camps and field hospitals; and on the 15th of March, without further delay, informed Marshal Pelissier upon the sanitary state of the army. The first question that I proposed to investigate was—whether the typhus raged only in the field hospitals, or whether it had also spread in the regiments. I became satisfied that the latter was but too truly the case, and gave directions that they should watch with the most scrupulous care, so as not to leave, either in the regimental infirmaries or the tents, any men tainted with the typhus;—on the first premonitory symptoms, they should be placed in the field hospitals. The human miasm of this disease does not become contagious until after the disease has continued several days, and especially during the period of the critical sweats;* and for this reason, the recommendation here given becomes of the highest importance. I also demanded that they should change

---

* The correctness of this point of theory is confirmed by the views of Hildebrand, and was in conformity with the facts observed in all the regiments in the Crimea.

the sites of all the camps from a soil that had become deeply impregnated with impurities; and that as often as the weather permitted, they should take down their tents, or at least raise the curtain that extended around the bottom to the height of a few inches from the ground. They might thus prevent the soldiers from remaining a great part of the day in their tents, which they kept hermetically closed, even in the finest weather. The floor of the tents, once dry, ought to receive a coat of white-wash, renewed from time to time, which would cleanse and harden the surface. The bed-clothing and garments of the sick should be hung up in the sun, as long a time each day as possible. Those which had been used by patients infected with the typhus ought to be exposed for several hours to chlorine fumigation before being again used.

A considerable number of regimental infirmaries had a defective arrangement, and, in place of two barracks, several had but one. The ground was not always protected against humidity, and camp bedsteads, or at least boards, were not always furnished to keep the sick from the ground. The interior of the barracks should be whitewashed with quicklime, and the floor and walls exposed to fumigations at frequent intervals. The rations should be increased a sixth part in the article of preserved meats, and a supplementary ration of wine should be issued daily, to give the army a greater power of resistance against the attacks of the malady.

I also recommended, as an excellent auxiliary to good hygienic condition, that exercise should be taken, to a judicious extent, when the weather was fine; for nothing is so pernicious as perfect repose, and idleness enervates both body and mind. The six thousand mattresses distributed four months previously, under the direction of the Intendant-general, were, in part, out of condition for service. There remained of them about 2,500, more or less. There was scarcely barrack room for more than 4,500 sick. The blankets were sufficiently numerous, but almost all of them were tainted; sheets and hospital clothes were wanting, as were also proper means for washing.

To supply these resources, which are so quickly expended in a campaign, we were obliged to overcome very great difficulties, in a country stripped of everything. The last ten days of February, the 20th to 29th, showed 519 sick discharged from the field hospitals cured, and 873 had died. In continuing the comparative inquiry, so as to include only those tainted with typhus, we shall observe a still more terrible result; for we have had in this disease only 27 recoveries to 385 deaths, while typhus, under ordinary conditions, does not carry off more than a sixth part of the number sick. Thus, at Constantinople, of 422 hospital attendants attacked with this disease in our hospitals, only 42 died.

In consequence of these facts, I proposed that we should retain in the Crimea only those sick with typhus, and that we should send all the others to Constantinople; and as the latter were the most numerous, their departure would produce an immediate disencumbrance, and allow us to use all our resources for the benefit of our unfortunate typhic patients. The latter, being retained in the Crimea, could no longer spread contagion in the ships and among the hospitals at Constantinople.*

Two hours after my sending in this report, Marshal Pelissier replied: "I have ordered that all your directions shall be carried immediately into effect in the regiments and field hospitals."

At the same time, strong encouragement came to me from France. On the 15th of March, the Minister of War wrote: "I await with much anxiety the news of your sanitary condition. Tell your associates in the health service, that I *thank* them. This word expresses the whole. The Emperor is informed of the new proofs they have shown of their zeal, their courage, and their self-denial. He has always relied upon these officers; but his faith in their devotion has been increased by a knowledge of the energy they have shown upon this occasion. I send to your assistance some sisters of charity, 200 infirmary attendants, and 20 aides. Can such a supply relieve you? At Marseilles and Toulon, there

---

* See in the Appendix the original report.

is much excitement, but nothing serious has occurred, although there is great apprehension. We are turning to account the good and prudent arrangements taken by you in your tour in Provence. The Emperor wrote to me this morning, and speaking of the sanitary condition of the army, he added : 'It is necessary to establish as soon as possible the field hospitals in barracks, which M. Baudens has advised, and let pressing orders be given accordingly.' I cannot do better than report the very words of the Emperor. I have addressed General Larchey by telegraph and letter, and have ordered that as many as possible of the sick shall be placed at Maslak. I have told him to arrange with the physicians, outside of all the pre-existing regulations, relative to the diet of the sick. He has full powers, and I will approve everything he may do. The Russian prisoners on the island of Prinkipo are in perfect health. I think that after their departure from Russia, if that event has not already occurred, a fine field-hospital might be established there. I have given orders to allow an additional pay to the physicians of one hundred francs per month.* I conclude by renewing the recommendation to retain at Constantinople all the sick whose removal is not demanded by want of space or by the absence of sanitary necessities." The Director of the War Department, M. Darrican, wrote to me: "Your position is terrible. We will do everything possible to relieve it."

On the same day, the President of the Council of Health in the Army addressed to me the following letter:—†

"MY VERY DEAR AND HONORED COLLEAGUE—The Council of Health has, with the greatest interest, taken into consideration the dispatches which you have suc-

---

* The radical decree of 1852 having abolished the extra war allowances to the medical staff of the army, privations that endangered their health were the result. The Minister of War, whose attention I had called to this pernicious condition of affairs, readily modified the evil growing out of the decree, by ordering a supplementary allowance.—AUTHOR'S NOTE.

† The Council of Health consists of three or five Inspectors, designated annually by the Minister of War.—TR.

cessively addressed to it, as also your last reports to the Minister upon the general aspect of the medical service at Constantinople.

"The Council hopes that the hygienic and prophylactic measures which you have proposed, and whose execution you have yourself superintended, will be followed by the favorable results that you anticipate, and that they will contribute, together with the reforms which you announce in the hospital rations, to the prevention of the development of infectious diseases, and to the arrest of their progress. The Council follows you in your labors with very great interest, and earnestly prays that the insufficiency in the number of the medical officers may not prevent the sick from receiving those assiduous attentions that may be so necessary for them.

"The Council, deeply touched by the death of so great a number of its courageous fellow laborers, returns you its thanks for having expressed over their graves in its behalf the just regrets which their loss has inspired, and desires you to be the medium for expressing to all the health officers of the army of the East the sentiments which are impressed upon it by their zeal, devotion, and self-denial in the midst of great dangers."

On the 16th of March, Marshal Pelissier decided that two field hospitals, which had become thoroughly infected, and which I advised to be abandoned, should be at once closed. The engineers immediately constructed two others, upon sites selected by me, upon high plateaux, placing the barracks twenty yards from one another, and the quarters of the physicians over two hundred yards from the hospitals. These two establishments have remained healthy, and have proved eminently useful. On the same day Marshal Pelissier ordered all the sick in the Crimea to be sent to Constantinople, except those sick with the typhus. I passed through the regiments, one after another; I conversed with the colonels, and imparted to them my observations. My counsels were everywhere listened to with eagerness, although not always followed with religious care. I could show that the mortality and sickness of the several regiments always bore an exact relation to the degree of solicitude which the colonels bestowed upon their soldiers.*

By the 28th of March, the good effects of these measures were easily observed, notwithstanding the protracted and severe winter. I informed Marshal Pélissier of this, drawing his attention also to some much regretted neglects, in the following report:—

"The measures which I recommended in my report of the 15th instant, for arresting the progress of typhus, and which your Excellency directed to be immediately carried into effect in the regiments and field hospitals, have borne their fruits.

"During the last ten days, the number of those entering the field hospitals shows a reduction of 500, as compared with those of the ten days previous, and the diseases are much less severe. The mortality in the Crimea has been reduced one-tenth. The cases of scurvy, on the 20th of March, in our field hospitals, amounted to only 649, in consequence of many being sent away; but at Constantinople we have in our hospitals, among 10,000 sick, 4,000 scorbutic cases. This disease, however, still fills, to a great degree, our regimental infirmaries; but is sensibly ameliorated since the distribution of fresh vegetables and the return of the sun. Some anti-scorbutic lemon juice, prepared according to the English method, has been placed at the disposal of the physicians in the army, and I look for good results from it.

---

* A general in the army of the Potomac, noted for the care which he bestows upon the welfare of his command, made a somewhat similar remark to the editor of this volume, upon the occasion of a visit for purposes of sanitary inspection, in March, 1862. He stated that his observations and inquiries had convinced him that the sickness of a regiment bore an inverse ratio to the efficiency of its discipline; and produced the records of his surgeons to prove, that for months the sickness had been greatest in those regiments which he ranked as the worst officered. All experience tends to prove, that without strict rules, well enforced, soldiers will not attend to those details of neatness of the person, the quarters, and the camp, so essential to health. In no way does the energetic disciplinarian appear to greater advantage than in this.—TR.

"So much for the general conditions. Let us now examine the state of the typhus.

"This disease has always one foot in the camp and the other in the field hospital. It has not gained ground; but still it must be acknowledged that it retains its hold, and menaces the army. We can safely assert that there are now a greater number of cures; medicine is, without doubt, favorable in some conditions, and its impotency less marked. Thus, during the last eleven days, we have counted 283 recoveries, to 699 deaths. The sum total is not much increased, as it was 1,457 on the 16th of March, and 1,606 on the 26th. It should be remarked, that since the 17th of this month, we have not sent to Constantinople a single person tainted with typhus.

"Two hundred and eighty-three cures in eleven days is evidently a very good comparative result, when we remember, that since the 1st of January, each period of ten days showed that the number of soldiers cured was 7, 14, 25, 36, 27, 62, and 45 respectively; but still this figure of 283, placed opposite to a mortality of 699 deaths, is very distressing. It shows conclusively that we should redouble our efforts to obtain a vigorous execution of every prophylactic measure recommended to the regiments and field hospitals. But permit me to inform your Excellency, that in passing among the regiments, I daily observe that many of the tents are not ventilated, that the clothing is rarely spread in the sun, and that the ground has not yet received the sprinkling of lime ordered. I either find the men absent, and the tents carefully closed, with large stoves, so as to be air-tight, or they are squat smoking within them, like Laplanders, exposed, without their knowing it, to the danger of reciprocal miasmatic poisoning; and yet, during the last fifteen days, the sun has not failed to vivify the earth daily with his presence.

"In the field hospitals, the measures directed are not executed as rapidly as I could wish, but still progress is made. The isolation of typhic patients has become almost complete, the bedsteads have arrived, and disinfection by chlorine fumigation is becoming general. Two field hospitals will be abandoned, according to your orders, on account of infection, as soon as those which your Excellency ordered are finished.

"I regard it as a most urgent measure, that the non-typhic patients should be immediately sent off. At least fifty cases of typhus show themselves daily in our field hospitals, among those entered for other diseases. This amounts to 1,500 cases per month, of whom it is shown, by actual results, that 1,000 prove fatal. It would therefore be 1,500 per month to strike out, if in our own field hospitals there were only sick belonging to this category. We should establish small hospitals in tents, at least three hundred yards from the actual field hospitals, for the temporary reception of new non-typhic cases which ought to be sent away. The result of this measure would be to keep up the morale of the soldiers, who have a horror of field hospitals and the contagion of typhus.

"Whenever the day arrives in which we shall have to treat only typhic cases coming from the corps, our situation will be materially alleviated. At the lowest estimate, the corps furnishes us daily seventy men infected with the epidemic; the recoveries and the deaths would very nearly balance this number, even in the present conditions.

"I have already had the honor of showing your Excellency that by retaining in the Crimea all those men infected with the typhus, we should relieve the fleet and the hospitals at Constantinople from the dangers of infection. I advised regimental physicians to redouble their vigilance to prevent any typhic patient from remaining in the regimental infirmaries. The only danger which threatens us is, the spread of the epidemic among the masses. It may be prevented by two different methods.

"1st. By continuing to remove from the corps, and sending to the field hospitals, those that show the first symptoms of the epidemic, with the object of arresting its propagation by infection.

"2d. In removing from the regiments the causes which spontaneously generate typhus. This may be done by causing to be put into effect the prophylactic measures laid down in the army orders, and remembering that the most important of these measures is the change of site of the camps from those places where the disease has been engendered.

"The urgency of this latter rule will appear more especially towards the time when the troops are to embark on their return to France. It will be necessary, at least fifteen days before that time, to make short marches with their shelter-tents, and change their encampments every two or three days. The country all along the ridge of the heights, which border upon the sea between Balaclava and Kamiesch, presents, in a hygienic point of view, favorable bivouacs for this purpose. I am well aware that these frequent movements must be irksome, and even vexatious, to those who do not see their importance; but the soldier, who alone with the physicians bears all the burdens of typhus, will, with them, heartily approve of it.

"If this advice is not followed, typhus will propagate itself on board ship; and as there will be no means of isolating the infected, and the epidemic might assume gigantic proportions, a heavy responsibility would rest on the physician who has not warned, and on the commander who heeds not the counsels given.

"The typhus will continue, so long as the causes that generate it remain; and the danger of not crushing it in the bud, and of letting it increase, is—that when once started, it develops itself rapidly by infection, and extends to the masses. Who does not remember the lamentable narratives of misfortunes occasioned by contagious typhus in armies?

"It is because I am thoroughly convinced that these misfortunes will not occur, if we seek to avoid them, that I insist so much—not so much to enlighten your Excellency, whose great wisdom and experience are known throughout the army, as to convince these men, whom I see ignorant of the most elementary principles of hygiene, living careless of the present, and, as it were, wrapping themselves up in a kind of Mahomedan fatalism.

"Are we to be frightened by it? Certainly not. Fear is a bad counsellor; but we must watch and foresee. By so doing, we are certain never to be unprepared, and we command the situation, whatever it may be. Supposing the typhus did not exist, there would still be advantage in religiously observing the hygienic rules which insure the health of armies.

"As for myself, I am reassured; because I have confidence in the enlightened and active concurrence of the commanders, and in the wisdom of the measures already adopted by the Minister of War. The personal service of the medical department continues to sustain losses daily, but its courage and self-denial increase with the emergency. I have taken, with the Intendant-general, certain measures, of a tendency to modify this state of things, which grieves me seriously. I intend to return with the mail of the 5th of April to Constantinople, where my presence may be of more use than in the Crimea; but before my departure, I respectfully beg your Excellency to give your advice, which shall be my command."

On the 30th of March, the Marshal replied:—

"Since the measures recommended to be put into execution, in your report of the 15th instant, have not been performed with all the care that was desirable in the several corps, I will again call the attention of the general officers to them. I will also carry into execution, as soon as the weather permits, your suggestions relative to encamping the troops for some time upon new grounds, and under the shelter-tents. As for your departure for Constantinople, you are a better judge than myself of its expediency; and I leave you entirely free to embark by the mail of April 5th, if you deem your presence more useful at Constantinople than in the Crimea."

An improved condition of affairs appeared, and on the 5th of April the Minister of War wrote to me as follows:—

"MY DEAR DOCTOR—I do not thank you again for the care which you have taken, for the zeal which you display in the interests of our poor sick; it would only be repetition. I agree with you that we should hasten to remove our unhealthy camps, placed upon grounds infected with miasms, and go as soon as possible to the heights that you have mentioned. I count upon the solicitude of Marshal Pelissier to give at once orders to this effect.* We are much engaged with matters relating to the return of the army to France. The numerous cases of typhus which have appeared at Marseilles, at Toulon, and on board our ships, are of a nature to cause us serious reflections. I have pointed out to General Rostolan that we have at the Island of St. Marguerite a hospital for four or five hundred sick, and also, under barracks or tents, beds for four or five thousand men. At Frioul, where we have already a hospital, I shall establish a camp for four or five thousand men. Lastly, in one of the islands of Hyères, on the peninsula of Gyen, I am establishing a third hospital and three camps, for ten or twelve thousand.

"Our ships will disembark their sick or healthy at one or another of these establishments.

"The healthy will remain eight or ten days, longer if necessary, in camp; they will walk about, bathe, will be well fed, will see the coasts of France; and, finally, all the conditions for restoration to health will be, as far as possible, combined. After undergoing this kind of quarantine, those who have sustained the test may be taken to Marseilles or Toulon, and sent to their proper garrisons. These are the measures which are being studied at this time. The results of the inspection you have made, in the localities that I have indicated, are very useful to us."

* We know that the Marshal only awaited the return of fine weather, which would dry the ground deeply soaked by the winter rains.

CHAPTER IV.

THE RETURN OF THE ARMY.

PEACE at length came, putting an end to our miseries. The relations between the allied armies and the Russians were soon established upon a very cordial footing, and on each side they celebrated, with fraternal libations, the termination of their long-continued sufferings. The Russians, French, English, and Sardinians, were to be seen walking arm in arm, singing and dancing, assisting each other to walk when they had taken a glass too much, as often happened, and giving them a hospitable couch for the night, whenever unsteady limbs rendered it impossible for their visitors to depart. The Russian General, Commander-in-Chief of the division encamped near Belbec, in speaking of this, said to me: "We have had some Zouaves for several days in our camps, who agree perfectly with our soldiers; by the aid of a very simple pantomime, they understand one another wonderfully, and drink together gaily. These Zouaves expect to be punished upon their return to camp, and have asked me for a certificate, showing that they were so well received that it was impossible for them to return to their regiment."

Steeple chases and military fêtes were held in the valley of the Tchernaïa, and the Arabian horses sustained their ancient reputation. In 1856, as in 1855, they resisted the rigors of winter and the hardships of the bivouac much better than horses of other breeds, thus justifying the assertions of General Dumas.* The races attracted large crowds, the soldiers attended without arms, and these reunions created a happy diversion of the spirits, which were depressed with apprehensions of the typhus. Moreover, dramatic artists from France

---

* It will be remembered, that in different essays printed in the *Revue des Deux Mondes* (numbers for December 1, 1851, and May 15, 1855), General Dumas first advocated the advantages of Arabian horses for the purposes of war.

gave very well attended exhibitions, every evening, in the theatre at Kamiesch, which prompted the formation of rival companies among the soldiers themselves. They compared the merits of the young chief actress at Kamiesch with those of a young Zouave bugler, and the opinions of critics were much divided. Had not most of the lyric heroes of the latter been slain in the storming of the Malakoff, it was thought that the theatre at Kamiesch could never have competed with that of the Zouaves.

In the bivouacs established upon the plains of Fedouchine, an immense ball-room was made, where figured the grand dames lately made wealthy in the villages of Pickpocketville and Rogueville.

Before leaving the Crimea, I visited once more, in company with Sir John Hall, the hospitals of our allies, and learned with certainty that the typhus had not reappeared since 1855. In the port of Balaclava, I visited an English steam frigate, fitted up as a hospital; it was arranged like a large ward, with 300 beds. They had carried attention to the wants of their sick so far as to place three or four cows on board, so as to have fresh milk during the passage. I inquired of the commandant how many troops a vessel of this size could transport, and he replied, "700 English, or 1,500 French, because the latter will quarter themselves anywhere, either on the deck, or between decks." The care which the English take of their soldiers, reminds me of the expression often used—"the English soldier is capital." This does not exclude from them—far from it—sentiments of humanity, but it attaches an idea of value, which should be preserved. On another occasion, when they captured at Balaclava a Russian officer and his family, an English general said, "Here is a good *Bank Note.*" The French navy had also some steam frigates transformed into hospitals; but the transportation of the sick was done by common merchant steamers or sailing vessels. The ships of the imperial postal service were particularly useful in this capacity. The sick had each of them a little mattress and a blanket.

From a general statistical account, published in

England by the Minister of War, Lord Panmure, it appears that from the 19th of September, 1854, to the 28th of September, 1855, the English army had 188 officers and 1,775 soldiers killed. There died of wounds, during the same period, 51 officers, and 1,548 soldiers; of cholera, 35 officers, and 4,244 soldiers; and of other diseases, up to the end of December, 1855, 26 officers, and 11,425 men.* These severe trials did not recur, and during the year 1856, the sanitary condition of the English army was in a very satisfactory state, even in the depth of winter.† To be convinced of this, it is only

---

* Since the above period, and up to March 31, 1856, 322 soldiers have died of wounds and diseases, making a total of 270 officers and 19,314 men. In addition to these, 2,873 soldiers were frozen, making a total of 22,457 deaths.

During the winter of 1854-5 (from the 1st of October to the 30th of April, a period of seven months), the mortality in the English army was distributed as follows:—

Infantry,	39	deaths per 100 men in seven months.			
Artillery,	18	"	"	"	"
Cavalry,	15	"	"	"	"
Naval brigade,	4	"	"	"	"
Officers of all arms,	6	"	"	"	"

AUTHOR'S NOTE.

† At an early period in the war, the English army had suffered considerably from disease; but this was remedied by a vigorous administration towards its close. During its occupation of Bulgaria, in the early part of 1854, cholera, diarrhœa, and dysentery made great havoc. The troops arrived before Sebastopol, debilitated by these diseases and the influences that had caused them, and were for weeks without tents or changes of clothing. The supplies arriving at a later period, left few causes of disease beyond those incident to the climate and the labors and exposures of the men.

Of fevers (not typhus), 30,376 cases and 3,161 deaths occurred. Their greatest prevalence was in March, 1855, when they gave 21 per cent. of the mortality of that month. The kind of fever that most prevailed in the English army partook of the character of typhoid fever, as seen in large cities, and showed an absence of great febrile action—a rapid prostration, slow convalescence, and tendency to relapse. It was often followed by dysentery, and the intestinal lesions which attended it occasioned secondary diseases that often proved fatal, especially if the patient exposed himself to irregularities of diet, or the use of intoxicating drinks, during the period of apparent convalescence. It was not uncommon for the febrile symptoms to disappear entirely, and prospects of health to gradually return, so as to admit of discharge from the hospital. But at a period more or less remote, a fresh inva-

necessary to read the weekly bulletin given to me by Sir John Hall, upon the day that I paid him a visit, March 25th, 1856.

Weekly Report, from the 16th to the 22d of March.

Effective strength of the English army in the Crimea and at Constantinople	70,042
Entered the hospitals and regimental infirmaries	1,883
Sent off, or discharged cured	1,641
Died	25
Remaining in the general and field hospitals	4,267

Proportion of sick to effective strength ......6.09 per cent.
Proportion of deaths to effective strength....0.03 per cent.

The return of fine weather dried the soil of the Crimea, and permitted us at length to remove the location of the camps to new and non-infected ground. The necessities of the war did not require, after the treaty of March 30th, any of the regiments to remain and hold the military positions of the left bank of the Tchernaïa, with its miasmatic emanations; Marshal Pelissier issued orders to abandon the old bivouacs, and remove three leagues to the south, upon the high plateaux, ventilated by the sea breeze, and inclining from the monastery of St. George towards Kamiesch. All of the barracks and large tents, contaminated by prolonged habitation, were replaced by Marshal Bugeaud's little shelter-tents. The sites of the camps were frequently changed, and these migrations every time effected an improvement in the health of the troops. Such removals often excited remonstrances on the part of officers, who were continually disturbed in

---

sion of the disease was very apt to occur; and after an uncertain continuance, with favorable remissions, death would in three or four months, if not sooner, close the scene. This fever was known as the "Crimean fever," and was never absent from the English army during the campaign.

Typhus fever proper, which made such frightful havoc in the French army, occurred among the English troops in 828 cases, of which 285 were fatal. It prevailed chiefly in the spring of 1855, and presented the usual appearances of the disease in England. The Russians suffered from the typhus even more terribly than the French—TR.

their arrangements. The Marshal, however, took no notice of them; he cared only for the health of the soldiers. He presided at the embarkation of the troops, watching that no regiments should be embarked but those in which traces of typhus had not appeared for some weeks; and did not himself leave the soil of the Crimea until the last regiment had departed.

On the 8th of April, 1856, I left the Crimea. M. Scrive, the physician-in-chief, with an enlightened solicitude, superintended and put into vigorous execution the hygienic measures which I had advised. Twice a week he sent me a bulletin of the sanitary state of the army before Sebastopol. Upon my arrival at Constantinople, on the 11th of April, I found still in the hospital three or four thousand cases of scurvy, and of 8,315 sick, there were 1,379 typhic patients. The mortality during the last ten days had only been 620, showing a reduction of more than a third upon the numbers for the preceding period of ten days. The Russian prisoners having just left the island of Prinkipo, I visited it on the day after my arrival, and found a perfectly salubrious site, which the Russians had not occupied. It was easy to establish there three hundred tents upon the border of a pine wood, and I designed sending thither a part of the scorbutic inmates of our hospitals. At the rate of six men to a tent, this would accommodate 1,800 patients, and I laid the plan before Marshal Pélissier. "The arrangement," said I, "would be simple; some boards laid together a few inches from the ground would serve as a floor and camp bed, and some mattresses, or, if these are wanting, some straw beds, with pillows, bed clothing, and blankets, will complete the furniture of this temporary establishment. The country air, now freshened with vegetation; the liberty of running in the woods, and of fishing on the sea-shore; and a diet largely vegetable, would yield results which we must never expect in our hospitals. I estimate that we could send monthly to Prinkipo 1,800 new scorbutic cases. We will not send the convalescents to France until they have undergone a month of this salutary quarantine;—thus easily and certainly escaping the invasion of typhus upon the voyage. Thanks to this increase of our resources, we may insure, without further interruption, the indispensable separation of typhic and non-typhic patients. We might intrust the attendance of these 1,800 scorbutic cases to the eight Crimean physicians who are at this time recovering from typhus at the convalescent hospital prepared for them at Sebastopol. If we persist in the salutary measure of sending to Constantinople all the sick not infected with the typhus, the personal attendance of the medical service of the Crimea, although so reduced by the epidemic, will nevertheless be more than sufficient, and a portion can be spared to return to Constantinople. As for the eight convalescent physicians, they will not, for a long time to come, be exposed to new dangers from the infection of typhus. At Prinkipo, in making the tour of visitation, they would regain their health at the same time as they were restoring it to their scorbutic patients."

On the 14th of April, the Marshal replied:—"I approve the putting into execution of this project, and by this mail give the necessary instructions." I at once went to Prinkipo, to prepare a large field hospital for 1,800 scurvy patients; and, thanks to the energy of General Pariset, who succeeded General Larchey in the command of the post at Constantinople, I completed this task in two or three days. Prinkipo took the place of Mytilene. Scarcely had the sick arrived, than they recovered their health, and were soon wandering over the island, elated and joyous.

On my way to Prinkipo, I stopped at Calchi, a neighboring island, where there was one of the two hospitals assigned to the naval service, the other being on the Bosphorus. Both were very admirably conducted by clever physicians. In front of Calchi there lay at anchor four or five large ships of war, which bore the yellow flag of the quarantine. They had been infected by the typhus in consequence of having transported the sick from the Crimea. A part of their crews, affected with this disease, had been landed, and were very well provided for, in immense rooms converted into hospitals, or in double tents. Immediately on my return to Constantinople, I effected the isolation of all the sick infected with the typhus, and a daily renewal of the bed clothing; the number of new cases appearing within the twenty-four hours fell at once to half the former number. Thanks to the mildness of the temperature, we were able to put in tents the typhic patients belonging to the hospitals *extra-muros*, thus securing a complete isolation and constant ventilation. The ships had still among their crews a number of typhic cases; and on the 14th of April, fourteen sailors, who belonged to a ship that had been selected to transport 300 convalescent sick, were sent to the hospital. The order for its departure was at once countermanded. I directed, in all the ships freighted by the government, an active surveillance to be instituted, and thorough hygienic and disinfecting measures to be adopted. I conferred with Dr. Fauvel, quarantine physician, who stated, that he had no directions to add to those which I had advised and caused to be put into effect. If the measures begun were thoroughly carried out, the typhus would speedily disappear. During the whole course of the epidemic, the population of Constantinople escaped the disease; nor did they evince any disquietude, being in this respect wiser than the inhabitants of the South of France, who were alarmed beyond measure at the importation of the pestilence through the arrival of typhic cases at Marseilles and Toulon. The Turkish government placed at our disposal, as auxiliaries, the most proficient students in their medical school, and the aid thus rendered gave us a very satisfactory idea of the organization of the medical corps of the Ottoman government. The director of the health service of the Turkish army was Thomal-Bey, a very important personage, and grand judge of Anatolia. This dignity corresponds with the rank of *Muchir*, or pacha of three tails. The generals of division are only pachas of two tails.

This high functionary is also Director of the School of Military Medicine, into which they admit students from civil life. He presides twice a week at the council, composed of the professors, and labors directly with the Minister of War. The sub-director of the school, Arif-

Bey, superintends the health service, and daily addresses a written report to the director. These health officers of the Turkish service, like the military staff of almost all other nations, have a line of grade and promotion corresponding with the officers of the army. In the Turkish army, all the chief physicians of the great hospitals have the rank of colonel, and their pay is even higher than that of those officers. The other physicians have the ranks of lieutenant-colonel, chief of battalion, and captain. The latter grade is borne only by a small number of the military medical officers.

In the last days of February, at the close of a conference upon the subject of typhus, at which the medical personnel of the hospitals of the military school attended, an English physician, Mr. Pinkoffs, who has distinguished himself by his great zeal for science, proposed to me an early convention of the English and Sardinian physicians. The idea at that time occurred to me to found a Medical Society, which should continue after our departure, by interesting in it the more eminent physicians of Constantinople and the professors of the Turkish Medical School, among whom was our learned countryman, M. Fauvel, quarantine surgeon. Mr. Pinkoffs seconded me willingly, and took all the necessary steps, which resulted, soon after our departure, in the foundation of a society, with an annual stipend from the Sultan, and called the *Imperial Medical Society*. I recur with pleasure to the time when, during my sojourn at Constantinople, I presided at this assembly of distinguished savans. The sessions of the new society were taken up by lectures and important discussions, and the medical press at Paris still continue to publish its transactions. As early as 1830, I had had the good fortune to revive in Algeria the course of studies of the Avicenne, the Razès, the Albucasis, &c., which had been interrupted for centuries; and it was with this idea that I co-operated in the formation of the first learned society in Stamboul.

Upon becoming assured of the good effect of the measures adopted to counteract the epidemic, happily on the decline, I was desirous of completing my investigations of the medical institutions of Turkey by a visit to their hospitals at Constantinople. The Ottoman government had reserved for its own use four military hospitals, located at the Séraskiérat, at Jeni Batché (new garden), at Gulhané, and at the Navy Office; the remainder having been generously placed at the disposal of the allied armies. The French occupied those on the European shore, and the English the fiue military edifices at Scutari, on the Asiatic side. All of these hospital establishments, built upon elevated and salubrious sites, and as much isolated from other habitations as a ship in quarantine, are a perfect acknowledgment of sanitary laws. A great number of openings admitted the air and light; the windows were double, to prevent the sick near them from being incommoded by currents of air through the joints, and to allow a genial temperature to be easily maintained in winter. Green curtains tempered the light, often too strong in this oriental region. On a level with the ground, openings made through the external walls ventilate the lower part of the rooms. The floors were painted in oil, or so thoroughly washed, that not a speck of dirt could be seen. Narrow strips of carpet, placed along the foot of the bed and the spaces between them, deadened the footsteps of the attendants. Smoking was strictly forbidden; convalescents could only smoke in the rooms specially reserved for this purpose. Excepting in some of the minor details, there was everything to praise. The beds appeared to me to be rather too near each other—the sick do not get air enough to breathe. This fault is in part obviated by the unexpected extreme cleanliness, and by the custom of keeping the doors and windows open. The mildness of the climate prevented those dangers which would have attended such a measure in France. The rooms are warmed in winter, and most of the windows open on to great inclosed galleries, where the temperature is never very low. Fumigations of chlorine, and especially of aromatic herbs, are repeated several times a day in all the rooms, carrying away the loathsome miasms generated by the sick—a practice which I would wish to see introduced into the hospitals in France, as it already has been into our general and field hospitals in the East.

The privies, placed in isolated sheds, communicate with the rooms by covered passages, and are paved with large blocks of marble kept always clean. A great number of small swinging windows, opening at the top, secured constant ventilation. Most of the bedsteads were of iron, and the mattresses, the clothing, and the blankets were of irreproachable cleanliness. Upon the night stand of each patient is placed a tray of tinned copper, holding a dish of drink, a plate, and a spit-box, so as not to soil the floors. These articles, which were of zinc, were kept very bright. A little board, fixed to the table, bore the number of the bed, the name of the patient, his disease, and the prescription of diet for the day. The apothecary wrote upon a separate memorandum book, at the dictation of the physician, the medicines which he was to administer. The hospital costume consisted of pantaloons, cap, slippers, &c.

The arrangements for washing would be an object of envy to our good Flemish housewives. Basins of white marble, attached to the walls, were, like our bathing tubs, furnished with pipes for supplying the water and letting it out after it had been used. The articles to be washed are soaked in large coppers, in which wood ashes* are macerated. In rainy weather, in winter, they exposed the linen upon driers, placed over large braziers. These prudent measures, so essential in a country frequently devastated by malignant epidemics, prevented the bad results of incumbrance and infection, so frequent in the hospitals of other countries.

The diet, appropriate to the wants of the East, is

---

* The custom of wringing clothing to deprive it of water is liable to tear it. The English, in their hospitals at Scutari, replaced wringing by drying in a turbine. The wet clothes were placed in a kind of circular metallic basin, pierced with many small holes, and this was made to revolve very swiftly upon its axis by means of a crank. The linen, obeying the centrifugal law, pressed strongly against the sides of the basin, where the openings gave exit to all the water contained. It would be an improvement to introduce it into our hospitals. The English call this ingenious device the "wringing machine."—AUTHOR'S NOTE.

healthy and simple. In Constantinople, beef not being much esteemed, mutton roasted, or served with vegetables, cabbage, chicory, or rice, forms the basis of hospital diet. The lighter form of diet most in use, is a cream of sweetened rice, of which the Turks prepare the national dish called *Malibi*. The sick take two repasts a day—one at eight in the morning, and the other at three or four o'clock in the afternoon. The first is composed of a simple porridge, thickened with rice, and the second includes, for a convalescent man, 300 grammes (0.57 lbs.) of cooked meat, some vegetables, and 600 grammes (1.14 lbs.) of bread. This allotment of food, too small for the first meal, and too abundant for the second, requires a better distribution.

The hospital at the Turkish Navy office presents a great luxury of arrangement, and, as a model, has no rival in any of the hospitals of Europe. In the little hospital of the palace of Bachistach, everything is in princely style,—rich carpets,—beds and curtains of silk, —choice supplies of food, and perfect attention to every want. Dr. Z——, one of the Sultan's physicians, who conducted me through the establishment, could not show me the apartments of the women of the harem, but informed me their principal disease was an ungovernable jealousy, continually aroused about things which to us would seem quite unimportant. They receive, from time to time, little gifts—for example, a box of sugarplums. The three or four hundred boxes must be exactly alike, or there ensue scenes of violence that endanger their health. Almost all of them die of pulmonary consumption at an early age. M. Z—— sent secretly to the more feeble of them some bottles of Bordeaux wine, to prolong their existence.

Henceforth, the grand and only object was the return of the army to France. The cases of typhus, already imported by our ships into Marseilles and Toulon, spread alarm among the inhabitants, and rendered great precautions necessary. The minister of war had, fortunately, taken the wise sanitary measures which were stated in a foregoing letter. On arriving in France, the army of the East had to undergo a kind of quarantine, in the hospitals and camps, established at the Island of Saint Marguerite, at Frioul, in the islands of Hyeres, and in the peninsula of Gyen, under the skilful medical direction of M. Maillot, inspector of the health service of the army. To establish the series of hospitals as stages on the route of the fleet, it was desirable that there should be a field hospital at Pirea, and another at Messina. The objections raised by the Neapolitan government prevented us from placing a hospital for typhic patients in Sicily. The ships, laden with troops, had orders to leave the infected cases at Gallipolis, Nazara, Malta, and Corsica, before their arrival in France. By debarking the sick men at each of these stations, it prevented contagion from spreading on board the transport vessels.

Two other sanitary stations were necessary, one between Nagara and Malta, and the other between Malta and Corsica. I went to Pirea and had an interview with Admiral Bouet Willaumez, and with M. Mercier, the French Minister. The Grecian Minister of Foreign Affairs, M. Rangabé, gave us immediately authority to establish a hospital, for typhic patients, in the island of Milo, which we went to examine. This island resembles, in form, a horseshoe, and only at the head of the harbor are there any low and marshy uninhabited grounds. The population, numbering about three thousand, live in villages perched upon the mountains. On the western side is Castro, where M. Brest, our Consul, a fine old man, to whom we are indebted for the Venus of Milo, resides. I selected a monastery, which had been abandoned since 1837, at the time when monastic property was reconstituted the domain of the Greek government. Knowing, by tradition, that the monks always selected for themselves the most salubrious and pleasant places, I made an ascent to this establishment by a winding but very practicable road for mules. I there found some large buildings, half in ruins, but which could be made serviceable—three or four fine kitchen gardens, and some beautiful level spots, shaded and perfectly adapted for pitching tents. An old centenarian and his family lived here, but occupied only one or two rooms. Water was abundant and of good quality. It was, however, difficult to install three hundred sick at Milo, and if infection had spread in the fleet during the voyage, this hospital would have soon proved insufficient. This fact decided us to sail for Candia, where the Sultan had given us permission to arrange a hospital establishment. We found in that island a fine well-aired plateau, which was reached by an easy mule road, which the Pacha promised should be immediately put in repair. Vely Pacha, formerly ambassador to Paris, placed at our disposal, one hundred officers' tents, for the formation of a hospital, which, fortunately, was not wanted.

The number of sick diminished rapidly in the Crimea and at Constantinople, the hospitals were emptied and closed, and, on the 6th of June, Marshal Pelissier wrote to me as follows:

"You will doubtless be recalled to France in a few days, and your mission to the army will then end. I will not let you leave Constantinople, without expressing to you my satisfaction at the manner in which you have discharged your duties, and without informing you that I have made a statement to the minister, of the useful and enlightened services which you have rendered to the army."

I left the East with the consciousness of having contributed, to the extent of my ability, to the alleviation of so many evils, and I may also say, of having witnessed the most appalling spectacles which have been seen for many a day. To the active instruments of death, which human genius had rendered so murderous, and which were never before collected in so great a number, within so small a space, were added the cholera, the scurvy, the dysentery, and the typhus. The constant and active solicitude of the government, the persevering efforts of the military administration, and the devotion of the medical staff, had ended, it is true, by conquering the epidemics, but at the cost of what sacrifices! If we consult the medical statistics of the hospital establishments, which should alone occupy our attention in this view, we would find that the number of deaths occurring in the field and general hospitals in the East, during the campaign, would amount to about 63,000, of which

31,000 occurred in the Crimea, and 32,000 at Constantinople.

An army needs moral excitement, to prevent homesickness and prostration. Religion exalted the troops of Godfrey of Bouillon; a spirit of chivalry animated the French officers at Fontenoy; the certainty of victory, maintained by a rapid succession of victories, sustained the armies of the Empire. It was a noble motive also which inspired our troops during the severe campaign of the Crimea; it was the sentiment of duty which animated our soldiers, without failing for a single day, in this struggle equally glorious against the enemy, and against privations and sufferings of every kind. Although other armies may have shown as much heroic ardor, impetuosity, and bravery, as the army of the East, none have surpassed it in stoicism, courage, and contempt of death.

APPENDIX.

I.

Economy in the Use of Linen for Dressings.

The Minister of War sent me, on the 30th of August, 1855, the following instructions:—

"At this time, when you are intrusted with the important service of Medical Inspector of the general and field hospitals of the Army of the East, I think it proper to send you some instructions relative to a thorough examination of the use and consumption of linen for dressings.

"The high price of this kind of linen, the scarcity of materials needed in its preparation, and the delays in its manufacture, render its adequate supply very difficult. It is therefore highly important, that the consumption of linen for dressings should be economized, in order that the surgeons may always be able to meet the more pressing demands of their profession. To enable you to render an exact account in this regard, I herewith send twenty copies of the instructions of May 12, 1845, and of the circular of January 15, 1853, relative to the use of these articles. These documents will give you full instructions as to the preparation of the linen and the means employed in supplying deficiencies. You will make such distribution of the papers as you may judge to be for the interest of your inspection.

"You are directed to satisfy yourself that the physicians in charge, in the establishments you may visit, use no more linen for dressings than may be absolutely necessary; and will request them to favor this end, as far as may be, by causing it to be washed and used again until completely worn out. You will be able to concert measures with the military intendants, Blanchot and Angot, for adopting the most effectual measures to attain this desirable result, through the concurrence of physicians and other accountable persons. In permanent establishments there should be no difficulty; those in charge can take advantage of the conveniences offered for washing. Experience has demonstrated that a pound of new linen may, by repeated washings, be made to go as far as six pounds. I cannot demand that this proportion shall be exactly attained; but an active supervision, and method, ought to yield results not hitherto reached.

"Your arrival in the Crimea will coincide with a measure which I think will be profitable to the treasury and facilitate the surgical service, without the sick having to suffer thereby. I will order, through the Military Journal, the employment of cotton, along with linen, for dressing, under such rules as may be furnished by the Council of Health. Large supplies of this linen will speedily be sent to the regular and field hospitals of the army, along with the compresses of carded cotton, of the use of which you are directed to explain the advantages. I charge you to use your influence among the acting surgeons to induce them to employ willingly these articles, with the view of reducing the demand for dressings within the limits of our means of supply. The important subjects brought to your notice by the present dispatch should be made the occasion of a detailed report in your labors of inspection. The military intendants, Blanchot and Angot, are informed of these instructions, and are directed to furnish all the means you may deem necessary to insure the execution of my orders."

II.

Alimentary Regulations of the Sick in the Sardinian Army.

THE following table shows the common rations of each meal (at ten in the morning and four in the afternoon), in the Sardinian army, and was furnished to me by M. Comizetti, chief physician of the Piedmontese army corps : it is here inserted as a document worthy of examination :—

	ounces bread.	ounces baked meat.	ounces pie or vermicel.	ounces rice.	ounces bread for soup.	pints broth.	pints wine.
Diet	—	—	1.32	1.76	3.52	0.48	—
Quarter ration ..	2.20	0.70	1.32	1.76	3.52	0.48	1.18
Half ration.....	4.40	1.40	1.32	1.76	3.52	0.48	0.36
Three-quar. ration	6.60	2.10	1.32	1.76	3.52	0.48	0.36
Ration*	8.80	2.80	1.32	1.76	3.52	0.48	0.36

The ration of meat for a day is only 0.55 lbs., while in the French hospitals it is double that quantity. It would not be enough to make a good broth ; but they add, like the English, the juice of preserved meats, called " essence of beef." The distribution of food commonly takes place twice a day, at ten in the morning and at four in the afternoon. Besides these rations, the patients receive, at eight o'clock in the morning, if prescribed by the physician, some soup, coffee, with or without milk, a milk soup, or some chocolate; but this distribution is not general; it extends only to those of the sick who appear to need it.

The rigid diet includes only broths and jelly broths. The common diet is composed of soups in the following order of gradation : a half soup, morning and evening, two, three, and even four soups, if the physicians prescribe them, or rather two soups, and coffee, alone or with milk, a milk soup, or chocolate. The soups of the

---

* The entire ration, after being ordered three days in succession, is, in exceptional cases, reduced to three-quarters, when they then return to the full ration.

diet are made at the option of the physicians, either with farinaceous ingredients, like vermicelli and semoulia, or of broken bread.

Besides the articles above enumerated, they have a number of kinds that they term extra-ordinary, which enter into the daily distribution, and will be mentioned when we come to notice the other elements of their rations.

The common quarter ration is composed of two or three soups, and the quantity of bread, meat, and wine stated in the table. With the quarter ration they may allow the patient the eight o'clock morning meal, with all the varieties above stated. With the quarter of the ration, the boiled meat may be replaced by chicken, veal, or mutton, or by vegetables, such as potatoes, spinach, carrots, cabbages, turnips, &c., according to tne season. They may also add two poached eggs, and 0.44 lbs. of cooked fruits or marmalade.

The two gills of *vin ordinaire*, usually added to the quarter ration, may be replaced by the same quantity of some generous wine, such as Bordeaux, Malaga, &c.

The half ration is composed of two soups, to which may be added for breakfast, at eight o'clock, another soup, coffee with milk, or chocolate, besides the quantity of bread, meat, and wine given in the table. The half ration of boiled meat may be also replaced by a half ration of ragout, fowl, gelatine, veal, or roast mutton, or by half a ration of vegetables; but it excludes dried fruits and marmalade, whose distribution here forms an exception, and is only allowed in certain particular cases. Wine may be substituted in the same manner as in the quarter ration. The half ration of meat may be divided into two quarters, as, for example, a quarter of ragout and a quarter of vegetables.

The three-quarter ration usually remains as in the table, but in exceptional cases, a soup, coffee, alone or with milk, chocolate, and the quantities of bread, meat, and wine, given in the table, may be added as breakfast. The three-quarter ration of boiled meat may be replaced only by a three-quarter ration of veal, mutton, ragout, vegetables, or salad, fowl and gelatine being excluded.

It may also be subdivided into a half ration of ragout, and a quarter of vegetables, or the converse.

The entire ration consists of the quantities of bread, meat, and wine, mentioned in the table. In the Crimea, in consequence of the inferior quality of the meat, and the special wants of the sick, it was permitted to add an extra ration of vegetables to that given in the table. Potato and fresh lettuce salads were, in consequence, often ordered for the scorbutic patients.

III.

Summary of immediate amputations performed in the Crimea; received and treated in the hospital of Gulhané, at Constantinople; and of delayed operations performed at the latter place, between the 1st of May and 31st of December, 1855.

IMMEDIATE OPERATIONS PERFORMED IN THE CRIMEA.

		Kind of Operations.	Number.	Recovered.	Died.
AMPUTATIONS IN THE BONE,	{	Arm,	91	51	40
		Forearm,	38	25	13
		Thigh,	74	30	44
		Leg, place of preference,	89	66	23
		—Submalleolar,	4	2	2
AMPUTATIONS AT THE JOINTS,	{	Scapulo-humeral,	25	16	9
		Radio-carpal,	5	4	1
		Carpo-metacarpal,	9	6	3
		Femoro-tibial, process of M. Baudens,	3	3	—
		Tarso-metatarsal,	7	5	2
RESECTIONS,	{	Head of the humerus, process of M. Baudens,	3	2	1
		Body of the humerus,	3	1	2
		Total,	351	211	140

CAUSES OF DEATH.

Purulent Infection,	41	Typhus,	6
Hospital Gangrene,	27	Pleuritic Effusion,	2
Gangrene,	20	" Abdominal,	2
Hæmorrhage,	2		
Chronic Diarrhœa,	36	Total,	140
Scurvy,	4		

DELAYED OPERATIONS PERFORMED IN THE HOSPITAL OF GULHANÉ.

	Kind of Operations.	Number.	Recovered.	Died.
AMPUTATIONS IN THE BONE,	Arm,	44	25	—
	Forearm,	14	9	—
	Thigh,	34	9	—
	Leg,	47	27	—
AMPUTATIONS AT THE JOINTS,	Scapulo-humeral,	3	2	—
	Coxo-femoral,	2	—	2
	Femoro-tibial,	5	—	5
RESECTIONS,	Body of the humerus,	1	1	—
	" " radius,	5	5	—
	" " femur,	1	—	—
LIGATURE OF ARTERIES,	Primitive carotid,	3	—	3
	Brachial,	6	4	2
	Femoral,	11	4	7
	Anterior tibial,	1	1	—
	Total,	177	87	90

CAUSES OF DEATH.

Purulent Infection,	29	Scurvy,		—
Hospital Gangrene,	16	Pleuritic Effusion,		3
Gangrene,	14	Chronic Diarrhœa,		22
Hæmorrhage,	1			
Typhoid Fever,	5		Total,	90

These tables show at once the great advantages in favor of amputations performed immediately after the wound, over those in which a considerable length of time has elapsed; and the results obtained at this hospital are similar to those at the other French hospitals at Constantinople.*

* The experience of the English army in the Crimea fully confirmed that of the French, in favor of early amputations from severe injuries by gunshot wounds, and showed that every hour of delay lessened the chance of a favorable issue. Dr. Macleod, in his "Notes on the Surgery of the Crimean War," has collected facts, showing that 1288 cases of the greater amputations, in various wars, from gunshot wounds, were followed by death in 396 instances, while 902 secondary amputations were followed by death in 586 cases. The per centage of 30·7 in the former, and 67·9 in the latter, nearly agrees with the experience of the Russians in the Crimean War. They lost one-third of the primary, and two thirds of the secondary amputations of this class. —TR.

The total of 140 deaths following immediate amputations, and 90 deaths to 87 recoveries in delayed operations, are excessively abnormal, and can only be explained by the striking deterioration of the health of the men at the time when they were wounded, and by the infection which prevailed in the general and field hospitals. When I had charge of the field hospitals of the expeditionary columns in Algeria, it often happened with me, as with my colleagues, that we did not lose a single patient from amputation. This was especially the case after the campaign of Mascara, when I obtained fifteen recoveries in succession. In one of the later expeditions in Kabylie, M. Bertherand, chief physician, met with results equally satisfactory in twenty amputations. A comparison of amputations with resections, shows also that the advantage rests, as I have often observed, in favor of the latter, thus proving that the domain of conservative surgery will be constantly enlarged. M. Thomas, the skilful surgeon whose practice I have above stated, coincides with my views, and I was much pleased, in examining his patients, to find the successful results obtained, in many cases, of wounds that appeared to demand an amputation of the thigh, through the use of my fracture box. Some of these cases will be given in the following memoir.

IV.

Observations upon several comminuted fractures of the thigh by bullets, resulting in cure when treated with Mr. Baudens' fracture box, reported by Mr. Thomas, Physician-in-chief.

BICHET, a sergeant of the 4th marines, of good constitution and sanguine temperament, was struck on the 7th of June, 1855, before Sebastopol, in his right thigh by a ball of large size. On his arrival at the hospital of Gulhané on the 18th, there were two openings, one on the outside and the other on the inside of the limb, at its lower third. The bone was broken with many splinters, and two of these 1.18 inches long, and of the size of a goose quill, with many of less size, were extracted from the place of entry of the ball, somewhat enlarged for the purpose. This reduced the case to one of much simplicity, and the limb was placed in a box, and subjected to extension and counter extension. In ninety-two days, the limb was consolidated without deformity or shortening, and without local or general accident. This subaltern officer rejoined his corps early in October.

DEMAIN, a fusilier of the 86th infantry, a young soldier of good constitution, was wounded June 9, in the right thigh, by the bursting of a shell, and arrived on the 18th at the hospital of Gulhané. He had a long wound in the upper and external part of his limb, with a comminuted fracture of the femur, about three quarters of an inch below the great trochanter. Many splinters were extracted, and the limb was placed upon cushions in the box, from whence it was taken three months after completely set, and with a shortening of only three quarters of an inch. The recovery, retarded by numerous abscesses, occasioned by splinters of bone and pieces of clothing left in the wound, was not complete before the beginning of October, when he was sent to France.

JUSEF-TESTANIEF, a Russian prisoner, received on the 7th of June, a fracture of the left femur below the great trochanter, and two wounds opposite that point, on the inner and upper part of the thigh, indicated the path of the ball which committed the injury. The finger reached a considerable number of large fragments, the removal of which, on the 20th of June, resulted in a loss of substance in the bone, estimated at an inch and a half. The limb was placed in a box, where it remains at the time of writing, having acquired so much firmness, that the patient can move it perfectly. The first wounds have closed, and pus flows only from two fistulous openings caused by abscesses from little splinters of bone, bits of clothing, and parts of the ball, of which the wound was never entirely freed. He was still in our wards, when, upon signing the peace, he returned to Russia, perfectly cured.

SAMOCHENKO, a Russian prisoner, received, June 7, a ball which passed from without inward, through the middle part of his left thigh, causing a comminuted fracture of the femur at that point. On the 19th, he was admitted into our hospital, where, after the extraction of four large fragments of bone and many pieces of clothing, his limb was placed in a box, and kept till October 20. The consolidation was then complete, but there existed a shortening of one and a half inches. Notwithstanding the restlessness of the patient, his cure was not delayed by any accident, either local or general. He is waiting in our wards, to be sent to Odessa.

ZEREPA (Simon), a Russian prisoner, on the 7th of June, was struck by a cylindro-conical ball, which passed through the right thigh, from front to rear, on the level of the great trochanter, fracturing the femur into splinters. On the 19th, when admitted at Gulhané, we extracted many splinters, and applied the fracture box. After a treatment of four and a half months, frequently embarrassed by abscesses of intermittent recurrence, and relapses of diarrhœa, this patient was sent to Odessa with a limb entirely solid, and shortened only three-fourths of an inch.

FERNET (Felix), sergeant-major in the 85th infantry, received September 8th, a ball, which entered the exterior and upper part of the right thigh, and fractured the middle of the femur. A few hours after, a field hospital surgeon extracted many splinters, and applied Scultet's bandage. On taking off this apparatus for the fourth time, on the 27th, upon entering the hospital at Gulhané, the limb was badly swollen, painful, and covered with a thick layer of bloody and putrid pus. It was at once placed in a fracture box, and covered with an emollient and narcotic decoction. On the 2d of October, the swelling having quite disappeared, we found that the ball had only made one opening, and that a cylindrical piece of bone four or five inches long, formed by the whole size of the femur, was detached and crowded towards the inner part of the thigh. The finger passed into the wound detected no other foreign substance. This condition being known and examined, we brought the fragments into position, and kept them in place by extension and counter extension, by the aid of adhesive straps, fastened upon different parts of the limb. Up to the 21st of October, several abscesses had opened, and the last one gave exit to the ball. Since the escape of this substance every day has witnessed an improvement in the condition of the limb, and the general health of the patient; and now (Jan. 4, 1856), the consolidation of fragments is far advanced, the limb preserves its form and proper length, the suppuration is slight and of good quality; the health of the patient is excellent, so that we may look forward to an early and complete recovery. Two months after Fernet returned to France perfectly cured.

These cases, I doubt not, will be studied with a lively interest by surgeons worthy of the name. They prove, contrary to received opinion, that comminuted fractures of the femur with wounds are not *necessarily* subjects for amputation, if we take care to extract the splinters and foreign bodies, so as to render a complicated wound simple, and then, by the aid of a proper apparatus, place the limb in a state of complete rest without shaking it, even during the dressings, and giving to the purulent fluids an easy outlet. I might have borrowed similar facts, from other surgeons in our hospitals at Constantinople, and thus increase the number, but they would have afforded nothing further of special interest. I cannot meanwhile pass in silence the observations recorded in the following chapter, and which attest the wonderful power of surgical art in skillful hands.

V.

Fractures of the femur, caused by projectiles. Cases of distorted bony union in broken thighs, remedied at long periods after the accident, by the use of the frac-

ture apparatus of M. Baudens. Report of M. Maupin, Chief Physician of the hospital at the Paradegrounds at Constantinople.

AMAR, a fusilier of the 21st regiment of the line, was shot in the left thigh, September 20, 1855. The ball, which was of large size, entered on the front exterior part, at the upper third of the limb, and lodged under the integuments, having passed through the great trochanter. The femur received a comminuted fracture, and considerable hæmorrhage occurred. The patient would not submit to the disarticulation of the limb at the hip. The wound was enlarged, the splinters of bone removed, and simple dressings applied twenty days. The limb, placed in a half bent position, rested upon a blanket, rolled up, so as to form a hollow trough. Some strips of bandage completed this rude apparatus, for which, in eight days, that of Scultet was substituted, and the patient was sent from the Crimea to Constantinople. Upon arriving at our hospital (October 14th), the wound was tainted with spots of hospital gangrene. The thigh, shortened at least five fingers' breadth, formed from without inward the arc of a circle, the heel touching the right calf, when the patient allowed his two thighs to take their own position. Amar is a man full of courage, and has faith that he will recover without an amputation. The limb was placed in an inclined plane, the foot being fastened to the foot of the bed. A cloth folded in many thicknesses, and placed between his thighs, effected counter-extension, but at the end of thirty days, no change in the deformity or shortening of the limb being apparent, the apparatus of M. Baudens was applied. On the 15th of March, when Amar was sent to France, the wounds, whose dressings were easily applied by aid of this apparatus, had been cicatrized for some time, the limb was slowly coming back, without torture, or even appreciable pain, and in an almost perfect manner, to its normal position, and the shortening was reduced, at the most, to three-fourths of an inch. The place of consolidation, very voluminous at first, was notably diminished, and its irregularities were lost in the general ossification. The thigh, which was much shrunk in November, had increased in volume, and its movements were in every way easy and free.

ALBARIC (JOHN), fusilier, of the 80th regiment of the line, September 8, 1855, received a musket shot on the outer side of the right thigh, at the point of junction of the upper external and middle third. There was but one opening. An examination at the field hospital of the second division, found fracture of the thigh, extracted several fragments of bone and the projectile, and applied a simple apparatus, consisting of compresses, bandages, and four splints.

On the 19th of October, Albaric was sent from the Crimea to the hospital at the parade-ground. The bony union had already become so firm, that it allowed of the limb being raised, but the natural curve of the thigh was much increased, and, moreover, at the place where the entry of the projectile was hidden by the cicatrix, there was a lump caused by the angle at meeting of the two pieces of the femur, and the limb was shortened full four fingers'-breadth. While the leg was slightly bent upon the thigh, the foot was *very much turned outward*. The limb was subjected to a very simple kind of continued extension, by being stretched between the head and foot of an iron bedstead, that is to say, it was mal-treated. The hospital gangrene devoured the cicatrix, and spread deep and wide.

On the 2d of January, 1856, the shortening and deformity of the limb were very nearly as they were on the 19th of October, and yet the solid appearance of the bony soldering had not suffered from the proximity to the hospital gangrene. The limb was placed in the permanent extension apparatus of M. Baudens, by the aid of which dressings may be daily applied to the gangrene with great facility. By the 15th of March, the limb had been gradually brought back in length, direction, and form, to correspond very nearly with the other, and the patient could move it as easily. Still the hospital gangrene, although carefully treated, remained. Twenty times it came to some kind of healing, and it would then relapse, showing, by its oscillations, the effects of the sick man's surroundings. At the place of the fracture, the thigh is now but slightly swollen.

SALARD, a grenadier, of the 18th regiment of the line, August 13th, 1855, met with a violent contusion of the right thigh by a musket ball, at a point three fingers'-breadth above the knee-pan; the integuments were only excoriated, but the femur was fractured, and an ecchymosis continued for several days. It will be proper to note that, four days before his wound, Salard left the infirmary of his corps, where he had been under treatment for arthritic rheumatism of the right knee, which had left a soreness and pain upon moving the joint. At the field hospital in the Crimea, the limb had been enveloped in three splints, resting upon three small cushions. In this condition, the patient arrived, on the 20th of August, at the hospital of the Parade ground. There was much swelling of the thigh and knee, which were extensively ecchymosed. The femur was fractured at its lower extremity, but the swelling did not allow of any examination into the condition of the bone. The limb was placed in a half flexed position, kept so by pieces of linen folded around it, and enveloped in compresses kept constantly wet with bran and water. The swelling being reduced, a fracture of the femur, just above the condyles, was found. The lower fragment was considerably raised, and the upper one made a lump very appreciable to the touch in the bend of the knee. The member was at once placed in the modified apparatus of the Hotel-Dieu, and retained nearly two months in it, with but little pain at first, but a great deal of inconvenience towards the end. From time to time the extension was increased, as much as an apparatus of this kind would allow of. The apparatus was removed in the early part of November. The bony union was then so far advanced, that the patient could raise the member at will, but the joining was very voluminous, and the limb was noticeably shortened. The knee was, besides, considerably swollen, and presented evidence of incipient hydrarthrosis.

Being impatient of further restraint, the patient rejected every measure proposed in view of relieving the difficulty, and attempted to walk. The swelling of the knee enlarged, it became painful, and the effusion increased. Salard was compelled to take to his bed again; but his constitution, which had hitherto been quite vigorous, became impaired in a very marked degree, and symptoms of a scorbutic character appeared. It became difficult, and then impossible, to raise the thigh without assistance, and it appeared probable that the bony union was disposed to yield to methodical and continued traction. The limb was placed in the apparatus of M. Baudens, where it remained about three months. By varied means, and especially by the aid of vesicatories repeatedly applied, the swelling of the knee was reduced to the proportions inseparable from a grave fracture. The shortening of the limb, which was more than three fingers' breadth, when taken from the Hôtel Dieu apparatus, had been extended to about a third of an inch of its proper length. The general projection of the soldered part was still perceptible in the bend of the knee, but deeper and better in the axis of the femur. Walking, though still difficult, was no longer painful. It grew stronger every day, and the knee, instead of swelling, became reduced to its proper volume. He left the hospital on the 15th of March, happy at having escaped by his second recovery, as far as was possible, the inconveniences that would have followed the first.

MEUNIER, a fusilier of the 21st of the line, September 8, 1855, was shot by a musket in the left thigh, the ball striking the superior external part of the limb, a little below the great trochanter, fracturing the femur at that point into splinters. There was no opening of exit, and no resulting circumstances occurred to indicate whether the projectile was buried in the limb or not. Eight hours after the accident, the patient was removed to the field hospital of the Second Division. The limb was then considerably swollen. Simple dressings were applied. The shortening amounted to nearly two inches. On the 20th of September, the apparatus of Scultet was applied. The wound, freed from splinters, tended rapidly towards cicatrization. Although hospital gangrene prevailed, he was not attacked with that disease. At the Hospital of the Parade Ground, where Meunier arrived on the 30th of September, the apparatus of Scultet was renewed, and the cicatrization of the wound was completed under this treatment. On the 28th of November, when the patient passed from the fourth to the first division of the wounded, the soldering of the bone was complete, although very voluminous and irregular, and had so far progressed, that he could move his thigh without appreciable pain. The shortening of the limb was three full fingers' breadth. The apparatus box of M. Baudens was adapted to the limb, and extension applied, both to the thigh and lower leg, and maintained till the end of February. On the 10th of March, the patient was sent to France. There was then scarcely any shortening perceptible upon measurement, and the bony knob at the point of union had well nigh disappeared.

This case proves anew, that when there is a faulty consolidation of recent occurrence, we may, notwithstanding the defective relation of the bony parts to one another, so far correct the misfortune, that in the least successful cases there will remain but a simple shortening compatible with the free use of the limb.

Up to what period may a faulty soldering of the bone be remedied? I am not, at present, able to show precisely, although my experience embraces nine successful cases. Still, I certainly would not accept the narrow limits which dogmatic teaching has prescribed in the application of trial methods; and I would observe, that several of my cases have been taken in hand within four months of the lesion, when the soldering presented such solidity that the patient could raise the limb. In these same cases, the most faulty union of bone yielded very readily to continued and graduated extension, and, two months later, had taken a proper direction, and showed an elongation that I had scarcely hoped to effect.

Of all the means or apparatus which I have used to correct, as much as possible, the deformities of the soldering, there has been none more convenient in application, or more sure of success, than that which I used in the cases here described. It combines all the modifications of traction and extension deemed necessary, and, at the same time, is as inflexible as we could desire in an apparatus of this kind. This action may be gradually increased without appreciable pain to the patient, conditionally, and here lies the whole secret of the success of all apparatus, that the different parts be carefully fitted, and the effects closely watched. As it does not cover up the limb, the proper dressings can be applied to the wound, without in any way interfering with its efficacy.

VI.

Observations collected by M. Beaufils, Aide-Major Physician.

I.—SPONTANEOUS LOSS OF BOTH FEET FROM CONGELATION.—CURE.

MANGIN, aged twenty-three, of feeble constitution and nervous temperament, belonging to the 62d regiment of the line, after five months' residence in the Crimea, had both feet frozen, while sleeping in his tent, on the night of March 19. He was sent to the hospital at Gulhané, where he arrived April 11, 1856. Both feet were cold and insensible, and the tissues had a black and lifeless aspect, with a line of separation around them at the ankles. The patient appeared weak and broken down. There was no fever. By the 30th of April, the circle of elimination had made progress, the tissues were deeply ulcerated, the tendons exposed, and fleshy granulations appeared along the edge of the wound towards the leg.

By the 5th of May, the bones of the legs were entirely denuded as far as the ankles, the left foot was only held by some tendons, and by the ligaments of the joint, which were completely dissected. A few clips of the scissors detached the foot from the leg, without a drop of blood following, and without its loss being felt by the patient. On the next day, May 6th, the same operation was performed upon the right foot. The patient was much pleased at being relieved from the annoying weight at the extremities, which had prevented him from turning in bed. On the 8th of May, he was placed under the influence of chloroform, and the operation of resection of the ankles was performed, as indicated by M. Baudens in his method of tibio-tarsal amputation. The borders of the wound were soon covered with fleshy granulations that completely enveloped the extremity of the bone, and now, May 30th, the wounds are almost entirely closed, the cicatrices are regular, and the patient, being considered as cured, is about to return to France.

II.—PARTIAL AMPUTATION OF BOTH FEET, IN CONSEQUENCE OF FREEZING, ONE COMPLETED SPONTANEOUSLY, AND THE OTHER BY SURGICAL AID.

MANAVELLA, of Piedmontese origin, a soldier of the second foreign legion, aged twenty years, of good constitution and nervous temperament, after fourteen months' residence in the Crimea, entered the field hospital of the fourth division on the 1st day of January, 1856, attacked by typhus. Being placed under a tent both his feet were frozen. Upon recovering from the typhus, he was sent to the hospital at Gulhané, where he arrived on the 14th of February.

The extremities of both feet, and as far up as the tarso-metatarsal articulation, were completely mummified; black, hard, resonant like a piece of wood, insensible, dry, and horny. The toes, being dried together, could not be moved. The general health was satisfactory, and the patient ate with relish the quarter, and after some days the half ration. The pain, which had been but slight, on the 20th of March became more acute. A circle of elimination was formed upon the right foot, and the affected part soon ulcerated. The suppuration became abundant, the pain very acute, and the slightest dressings could not be applied without exciting cries of distress. The extremity of the foot drew heavily upon the ligaments, causing great torture to the patient, who, nevertheless, could not bring himself to consent to an operation. On the 5th of April the medio-tarsal articulation opened, the ligaments were in part destroyed, and the skin held only to the bottom of the foot. A few clips of the scissors completed the amputation that had been begun and partly finished by nature. The scaphoid bone remained in place; the three cuneiform bones and the cuboid bone were removed, and but little blood followed the operation. The borders of the wound were covered with bright florid granulations, of healthy character.

On the 20th of April, the flesh covered the scaphoid bone. The granulations were of a vermillion red color, the wound was diminished in extent, the suppuration was of good quality and less abundant, and the pain had ceased. On the 30th of April the wound presented a very good appearance, and was much reduced, and now, May 30th, it has entirely closed.

The left foot took much longer time to eliminate itself than the right. Although the toes and lower parts of the foot were equally mummified, black, shrivelled, hard as wood, glued together, and completely insensible, yet the circle of separation had not begun to appear by the 20th of April. Thinking that, the gangrene being only superficial, vitality might be restored to the toes, the foot was subjected to a continued irrigation of cold water. On the 20th of April, the skin of the great toe appeared macerated and white in places. On the 24th, the extremity of the foot had passed from black to white. The heel and lower part of the leg were warm. The patient feels well; there is no pain or chills; the irrigation of the limb was continued. On the 26th a line of elimination appeared at the level of the tarso-metatarsal articulation; the tissues over the tarsal bones, which were slightly swollen, became white, and almost insensible to pressure. The toes were still mummified, hard, and covered with a thick layer of white epidermis, that could be removed like a blister by the aid of a spatula. On the 27th, the tissues appeared to take new life, and the great toe, particularly, showed a slight tinge of redness. The patient felt a little life in it. On the 30th, the foot presented the same favorable aspect, and the irrigation was constantly continued. On the 2d of May, the mummification of the tissues had almost entirely disappeared, the toes showed a fleshy color, especially the great toe and the third. They were less stiff, less glued together, and there was neither pain nor offensive odor. On the 3d of May the patient could stir the toes, and effect some slight movements. On the 4th of May, a line of demarcation appeared on the back of the foot, without inflammation, pain, or redness, and worms were continually multiplying in the wound. On the 6th of May it was deeper, and continued to extend, the patient, meanwhile, showing no signs of suffering or of reaction. The irrigation, which was continued, prevented pain and suppressed all bad odors. At length, on the 13th of May, the ulceration having continued to extend in depth, a considerable change in the osseous tissue could be detected. Mr. Thomas, Chief Physician, performed a medio-tarsal amputation, preserving as much as possible of the soft parts. A hæmorrhage occurred, which was arrested by the perchloride of iron. The patient, at the time of operation, was under the influence of chloroform. At this time, May 30th, the wound has completely closed, and the patient may be considered cured.

These two cases show very conclusively the appearance and progress of injuries caused by frost-bite. They indicate the prudent course to be followed by the surgeon, whose only object should be to second the efforts of nature, and to come to her aid only when she cannot effect her object alone. The aid of the surgeon should be particularly withheld when the constitution of the patient is broken down, or when he is infected with miasmatic poison. We have, in the foregoing pages, shown that operations, apparently the most innocent, were sufficient at times to excite hospital gangrene in the hospitals of the Crimea and at Constantinople. It was not till later, when the hospitals had been purified from this terrible scourge, that surgical operations were relieved from the saddening occurrence of reverses, and that amputations could be performed with fair chances of success.

Report of Mr. Thomas, Physician-in-chief, upon the Sanitary Condition of Constantinople, during the month of January, 1856, addressed to the Medical Inspector of the Army of the East.

" From the 1st to the 31st of January, inclusive, 13,520 patients were treated in our hospitals, of whom 895 were wounded by the fire of the enemy; 504 by other causes; 1,075 were frost-bitten, and 11,048 were fever patients received from the Crimea, from troops in passage, and from the garrisons. Since the first of January, the arrivals of troops sent for medical treatment from the Crimea, have been more frequent and numerous than in the preceding months, and the increasing numbers of sick under care, have rapidly consumed the means provided by the administration, who, counting upon the continuance of the sanitary condition of November, had discontinued the hospitals at the Preparatory School, at Maslak, No, 1, and reduced by a third, the number of beds in the other establishments.*

"Excepting some cases of old wounds, kept in the hospitals since the taking of Sebastopol, and a few hundred men frost-bitten, we find that those sent to the hospitals during the month, consisted entirely of febrile patients tainted with scurvy, diarrhœa or chronic dysentery, and typhus. Scurvy is, so to speak, the exclusive disease of the febrile patients arriving from the Crimea during the month, and exists with them all, either alone or complicated with other affections. Many of them present symptoms of advanced cachexy, great anemia, general swelling, fungous ulceration of the gums, sanguineous and painful infiltrations of the limbs, serous effusions in the cellular tissues, peritonitis, pleuritis, pulmonary œdema, diarrhœa almost always accompanied by hæmorrhages, etc.

"Cases of diarrhœa and chronic dysentery are very common, stubborn, and destructive. I have already, in several of my preceding reports, had the honor of stat-

---

* This latter measure had been adopted at my request, in view of disencumbering and purifying the hospitals.

ing to you, facts concerning a special complication which has for some months changed completely the appearance of our wards. It prevailed in our hospitals during the winter of 1855, and has been commonly denominated the typhic condition. During the present year, it has also prevailed, at the same epoch, and attended with almost identical symptoms. However, our physicians are far from agreeing as to its nature, and while some regard it as a variety of typhoid fever, others consider it a severe form of remittent fever; the majority regard it as the typhus of camps, hospitals, etc.

"This typhus threatens to assume the dimensions of a great epidemic, and is observed chiefly among the sick arriving from the Crimea, where our soldiers live under conditions most favorable for its development. They are there deprived of hygienic attentions, are overcrowded and exposed to the deleterious emanations of a soil profusely charged with organic matter of every kind, and impregnated with the effluvia of more than twenty thousand decaying men and animals. The residence in our wards of a certain number of men infected with this pernicious disease, has developed a true epidemic typhus, whose malign influence is reflected upon those sick of other and chronic diseases, and upon the attendants, sisters of charity, and physicians upon duty.

"This disease, notwithstanding its many points of resemblance to typhoid fever, appears to me to differ from it essentially, in the manner of its attack, the progress of its symptoms, the duration of the various attendant conditions, and the pathological changes that result from its action. Thus, contrary to what takes place among the greater number of cases of typhoid fever, we rarely have a premonitory period; and the greater number of cases of typhus which we have observed, attacked convalescents from other diseases, or persons in health.

"We have seen it almost always appear suddenly, with an initial chill, followed by a burning fever, and severe pain in the forehead, which was soon followed by a furious or tranquil delirium, according to the degree of nervous susceptibility, or perhaps the paludal intoxication of the patient. These first symptoms were followed closely by stupor and general prostration, and often this lethargy showed itself the first day in debilitated patients. These showed an extreme depression of vitality in the expression of the countenance, and an indescribably death-like cast of features, which constitute one of the pathological symptoms of typhus. Loss of hearing existed in a great majority of cases.

"During the whole course of the fever, the skin remained dry and burning. It was often covered with petechial spots, but rarely among the sick whose constitutions, impaired by previous diseases, had not strength to bring on reaction, and suscitate the stage of elimination. This eruption, either separate or confluent, and red, purple, or brown, appearing from the first to the third day, differed from the lenticular rosy spots of typhoid fever in their form, which resembled the eruption of measles, by their early appearance and by their duration, which was rarely more than four days.

"The thirst of the patients was proportioned to the intensity of the fever. Frequently the tongue remained clean and moist, and only in very prolonged and severe cases, or with those having a scorbutic taint, was it covered with a yellow or brown coat. We have at times seen it of a blood-red color, attended with extreme dryness, and a remarkable uncertainty in its motions.

"The bowels usually preserved their pliability and normal susceptibility; and if they were the seat of pain, it was transient, light, and spread over the whole surface of the abdomen, or fixed under the sub-umbilical region. It was rarely felt exclusively in the right iliac fossa, and crepitation existed in this region only in patients suffering from diarrhœa. Constipation is the more common condition with typhic patients, while meteorism, so common in typhoid fever, was very rarely observed.

"The lungs of scorbutic patients tainted with typhus often became the seat of congestions, which were announced by dyspnœa, a sense of oppression, and râles of various kinds. This pectoral complication of typhus was generally more grave and deadly in its effects than the abdominal and encephalic. Sometimes the pulse was full, hard, and frequent, especially at first; and at other times it was small, sharp, quick (a hundred and thirty to the minute), and often maintained itself in this condition till the end of the disease. In most instances it was weak and languid, and in grave cases, towards the close, it was thread-like, and scarcely perceptible. It however varied considerably, according to the constitution of the individual.

"Nasal hæmorrhages were common in cases of typhus, rarely appearing before the fourth day, and still more rarely showing themselves in the premonitory stage, when such occurred, or at the beginning of the disease, as in typhoid fever. The mean duration of this malady might perhaps be estimated at eight days; frequently it terminated at the end of the first week, and rarely went beyond the second. Grave cases of typhus, which were prolonged beyond this period, oftentimes relapsed into conditions very similar to those of typhoid fever. The mouth became dry, the tongue and gums were covered with sordes, the bowels were inflated, the right iliac fossa was painful and the seat of a manifest crepitation, and at length a diarrhœa appeared. At a still later period, coma supervened, with a general loss of sensibility and the functions of the senses, thus completing a striking resemblance between the two diseases. Frequently the typhus ended as suddenly as it began; in fact, it is not a rare thing to find the disease all at once become aggravated, and terminate in death in a few hours, while very severe cases would as suddenly amend, and return quickly to health. In a considerable number of cases of typhus, we have seen the malady attended with fetid sweats having an odor of rotten straw, and with intestinal fluxes and parotitis. These abscesses in the parotid glands have more than once modified the progress of the disease in a favorable manner, especially in the case of a young aide-major, M. Garny, whose recovery from a severe attack of typhus was not really determined until the appearance of an abscess of this kind.

"No matter what cause induced a favorable termination of this disease, the convalescent returned rapidly and steadily to health, thus presenting a contrast with the slow and almost fatal irregularity of typhoid fever. This difference in the progress of convalescence may be explained by the diversity of the intestinal lesions existing in the two diseases. Thus, in cases of typhus ending fatally, after an attack of a person in full health, we simply find along the intestinal canal some red and yellow discolorations in patches or bands, and numerous arborescent injections of the mucous membrane, a few pimples, and some of Peyer's glands, pointed or reticulated; a condition which we constantly find in a multitude of different diseases, and which cannot be regarded as a constant pathological character. We have never observed the glands honey-combed, fungous, pustulous, ulcerated, or gangrenous, nor the mesenteric ganglia engorged, nor any of the other special lesions of typhoid fever.

"This disease, of which we have enumerated the principal symptoms, has a strong tendency to spread from one patient to the next. Thus we have seen it, in our wards, communicate from the first patient to his nearest neighbors, and so on from them to the attendants, the sisters of charity, and the physicians. We have met it in persons whose age would be considered as a protection against an attack of typhoid fever, and with young men who are well known to have been attacked and to have been cured (Messrs. Lardy and Laval, Aides-Major).

"These facts should be considered sufficient to establish the nature of typhus, and prevent it from being confounded with typhoid fever. If, furthermore, it did not differ essentially from it—

"1st. In the manner of attack and progress, the absence of premonitory symptoms in almost all cases, and the sudden appearance of nervous symptoms, which rapidly reach their state of greatest intensity.

"2d. In the form, period, and duration of the exanthematous eruption.

"3d. In the period of the appearance of epistaxis.

" 4th. In the frequent absence of disturbances in the region of the bowels.

" 5th. In the shortness of duration.

" 6th. In the rapidity of convalescence.

" 7th. In the frequent occurrence of parotitis.

" 8th. In the difference of the pathological changes observed in the two diseases.

"This epidemic typhus, after being a long time stationary, has suddenly assumed very considerable proportions. During the first fifteen days of January, the number of patients tainted with the disease was not over a hundred, but now it has arisen to sixteen hundred. It has thus far spared the officers of the army; but, on the other hand, it has raged cruelly among the surgeons, the sisters of charity, the chaplains, and the attendants belonging to the field and general hospitals. Of the physicians treated for this disease in the hospital at the Russian embassy, eight are in a most serious and alarming condition. They are Messrs. Girard, de Santi; Millemberger, Arondel, Rioublanc, majors and aides-majors of the field hospitals; and Volage, principal, Cornac, major, and Bauchet, aide-major of the general hospitals. Since the last of January, we have had Messrs. Dulac, Savaëte, and Lardy. Several, whose names I do not know, succumbed to the disease at Eupatoria and Sebastopol.

" Your hygienic prescriptions to insure the salubrity of our hospitals have been carried into effect. At your instance, a depot of convalescents has been opened, and the new localities have been prepared for the accommodation of five thousand sick, with the view of scattering them, and thus prevent encumbrance in the existing hospitals. The food of the sick has been modified and adapted to the wants of the patients whom we treat; and there is reason to hope that these wise measures will eradicate the disease, and that we shall soon be able to announce an amelioration in our general condition of health."

VIII.

Instructions relative to the establishment of a Dépôt of Convalescents, in a part of the Camps at Maslak.

1. ORGANIZATION.—The convalescents shall be divided by sections and arms, as much as possible, the effective maximum being 200.

2. The sections shall be commanded and managed by lieutenants, or sub-lieutenants, having under them sub-officers and corporals. They shall each be taken from among the convalescents, or in default of these, from the 84th line, which regiment shall always furnish for each section, the sub-officer and corporal who shall perform the duties of sergeant-major and quartermaster.

3. A captain, selected from the 84th, shall exercise supervision over the officers of the section in all that relates to discipline, internal police, orderlies, etc., and shall sign for the various sections, such papers as orders for purchase, bills, etc. He shall be responsible to the council of the administration of the corps, with whom he is to confer directly, with regard to administrative affairs.

4. ADMINISTRATION.—The sections shall be placed under the general supervision of a superior officer, taken from the 84th, who shall direct the general service, secure a proper discipline, and employ the captain as his intermediary for transmission of his orders.

5. INTERIOR REGULATIONS.—Each section shall form an administrative unity, in the same manner as the companies of a corps.

6. The sections shall mess separately.

7. The mess expenses shall be the same as in the regular army corps. There shall be added a supplementary sum of five centimes (a cent) a day, which shall be devoted exclusively to the purchase of fresh or preserved vegetables.

8. There shall be allowed to each sub-officer and soldier present, including those charged with the com-

mand, a ration of wine daily, in addition to the regular ration of sugar and coffee.

9. The rations of provisions shall consist exclusively of bread, fresh meat, rice or dried vegetables, salt, sugar and coffee, at fixed prices, and the fixed allowance of fuel.

10. The additional ration of wine shall be distributed one half at each meal.

11. The ration of sugar and coffee shall be issued at about eight o'clock in the morning.

12. BARRACKS AND ENCAMPMENTS.—The convalescents shall be placed in barracks instead of tents, and these shall be furnished with stoves.

13. To each man there shall be allowed a mattress and a bolster, which shall be placed upon the camp beds.

14. In the absence of camp beds, trestle bedsteads of wood shall be furnished.

15. The men shall be provided with camp-blankets.

16. The utensils and other furniture for this service, will be furnished from the public stores upon the order and responsibility of the corps to which the men belong.

17. SANITARY SERVICE.—The sanitary service shall be intrusted to a physician of the grade of major, who shall be designated by the chief physician of the service of the hospitals at Constantinople. He will be assisted by one or more aides-major, according to the wants of the establishment.

18. The men shall be taken out to walk, according to the direction of the chief physician, who shall limit the time, and if need be, shall cause the feebler ones to be attended by an aide-major.

19. A sanitary visit shall be made weekly, and those who are deemed able to rejoin the service, will be sent to their corps. Pleasant days are to be selected, as far as practicable, for sending them off.

20. Such men as prove to be incapable of complete recovery at the depôt of convalescents, may, according to circumstances, be sent to the hospital, or be allowed sick leave, or be sent to the depôt of their corps. In these two latter cases, the major-physician charged with the duty, shall furnish certificates of visit, which visit

shall be made by the chief physician of the health service, in the presence of the Brigadier General.

21. The barracks shall be warmed only when the major-physician may deem it necessary.

22. One or more barracks shall be placed at the service of the major-physician, to be used as a regimental infirmary.

23. The medicines and dressings necessary for use in the infirmary, shall be supplied from the public stores, to be charged against the funds of the corps that may require them.

24. FORMATION OF THE DÉPÔT—*Sending to the Hospitals.*—The dépôt for convalescents shall be opened on the first of February next, and instructions shall be issued in time to enable the chief physicians of the hospital service, and other accountable officers, to select and send to Maslak such men as they may deem proper to be admitted. Those whose duty it is to report lists of names, shall indicate the number of the regiment, the Christian and family name, and the rank, and they shall be classified as infantry, cavalry, artillery and engineers, and other arms. These lists shall be addressed to the military intendant, on the 28th instant, and shall be at once sent to the general commanding the military division.

25. The hospitals shall send their men upon the first formation of the dépôt, in the following order:

February 1st. Pera, Dolma-Batché, Military School, and Parade-ground Hospitals.

February 2d. Gulhané and the University Hospitals.

February 3rd. Daoud-Pacha, Maltepé, and Kamis Tchifflick Hospitals.

26. The first formation shall embrace a section for each arm and, if need be, two from the Infantry,

27. The sections shall form a single series of numbers, and the designation of the arm shall be added to the number of each section.

28. After the first formation, the hospitals shall send only upon certain days of the week, viz:

Monday—Gulhané and the University Hospitals.

Wednesday—Pera and Dolma-Batché Hospitals.

Thursday—Military School and Parade-ground Hospitals.

Saturday—Daoud-Pacha, Maltepé, and Kamis-Tchifflick Hospitals.

In case of bad weather, the sending may be delayed till the next day, or to the day named in the week following.

29. The selections shall be made by the chief physicians, at least forty-eight hours before sending. Lists, including details mentioned in Article 24, and arranged according to the arms, shall be sent to the chief of the staff, on the day before the day designated for the removal, in order that the commander of the depôt may be prepared in time.

30. Such men as are known by the major-physician to be unable to walk the distance, may be sent in chairs or arabas, according as may be most convenient.

31. In all cases the Administration will take charge of the transportation of all property belonging to the men, sent to the depôt.

32. Each man sent by the hospitals to the depôt shall be furnished with a ticket of discharge, which shall be at once sent to the major-physician, and by him returned to the men when they are sent to the stations of their corps. The tickets of the men entered at the hospital or sent to France, shall be lodged with the captain performing the duties of military sub-intendant, who shall forward them to the smaller depôts of the corps, at the same time notifying the depôts of the present location of the men.

IX.

Report to Marshal Pelissier.

CRIMEA, GENERAL QUARTERS, *March* 15, 1856.

"TO HIS EXCELLENCY, THE MARSHAL:

"I have hastily passed through a part of the camps and field hospitals, and without further delay, consider it my duty to inform your Excellency of my opinion, as to the sanitary condition of the army.

"The first point which I wish to determine, was, whether typhus prevailed in the field hospitals only, or whether it was also prevalent in the regiments. To be convinced of the latter, it was only necessary to attend the morning visit when the men presented themselves at the regimental infirmaries. To the inexperienced eye, the premonitory symptoms of typhus may be confounded with those of other diseases, and some of our physicians did so mistake them at first, but in the hospital all doubt quickly vanishes. To the initial chills, with headache, succeeds a dulness and stupor, which distinguishes typhus as plainly as the symptoms of cholera, which can no longer be mistaken. The production of typhus in the midst of regiments, is a serious fact, and unhappily as well established as the fact of its propagation by infection. Two great indications present themselves here: first, prevent its spread among the masses; and second, remove the causes. The measures to prevent the spread of typhus among the troops, are simple, and easy to put in execution.

"It is only necessary to carefully watch that no patient infected with the disease shall be allowed to remain either in the tents or in the regimental infirmaries, and to send the soldiers, on the first appearance of the symptoms, to the field hospitals.

"The importance of this counsel will be appreciated when it is remembered that human miasm does not appear to be contagious until the expiration of several days of sickness, and above all, at the period of the critical sweats.

"The preventive measures to be applied in the regiments, are as follows, and observe, Marshal, that I shall only recommend such as are within our resources, and those which your Excellency, with wise forethought, has given to the army.

"Awaiting the return of fine and settled weather, to allow of changing the site of all the camps, the soil of which is deeply impregnated with impurities, it would be well, as often as the weather permits, to take down the tents, or at least to draw up the circular curtain as high as the safety cords will permit. This operation, which I have often witnessed, can be easily accomplished if the tent-pins are four or five inches above the ground. Six men occupied five or six minutes in completing it. Furthermore, recourse should be had to this hygienic measure, which serves to ventilate and dry the ground, always damp, if not actually muddy, whenever it does not rain or snow; and for a stronger reason, when the sun shines, and when a fine breeze is blowing.

"In order to secure the punctual execution of this prescription in the 15,000 to 20,000 tents, it should be indicated by drum-roll in the morning, when the men get up, and in the evening at the time of retiring. It would have the effect of preventing the soldiers from remaining the greater part of the day in their tents, which they keep hermetically closed, even in the finest weather—such as we have had for the last three days. It is quite enough to have them closed at night, and to be obliged to breathe a contaminated air during sleep.

"The ground of the tents, once dry, should receive a coat of lime, which would harden and purify it.

"The blankets and articles of clothing should be hung out in the sun, as soon as it appears, so as to dry out the dampness and purify them. Those which have been used by typhic patients should be subjected to several chlorine fumigations, before being again put to use.

"As soon as the weather permits, the shelter tents should be removed to another location. These tents have not only the advantage of affording to the inmates a sufficient protection from rain and atmospheric changes, but allow the air inside to be renewed by the passage of fresh air through the interstices of the fabric.

"A great number of the regimental infirmaries are badly arranged. Instead of two barracks, many have only one. The men are not always protected from contact with the damp ground by camp-bedsteads, or even planks. The barracks ought to be whitewashed inside and out, and often fumigated. Why not supply all of them with the little wooden frames, like those in the infirmary of the 81st of the line, which is a model of good arrangement, and those which are supplied to the sub-officers and to the soldiers of the guard? The air circulates freely underneath these little sort of litters, and when possible, they might be covered with a sack of hay or straw, as in the barrack of the 81st of the line, which provides a good bed. Since I have spoken of the 81st of the line, I will further say that I experienced great satisfaction at witnessing the fine hygienic arrangements in the tents of this regiment. Each soldier has a separate camp-bedstead and a mattress; cleanliness pervades everywhere, and we may ask why the care bestowed by the officers on the men in this instance, should not be emulated by all the others. The results of such a system are that the 81st have an effective strength of 2,400 men, ready at a moment's notice to encounter the fatigues of war.

"*Food.*—All the food, with the exception of the fresh beef, is good. The beef is flaccid and gluey; the animals are reduced to a state of marasmus when killed, and their flesh affords but slight nourishment; but this eventually has been anticipated by the solicitude of the government. In the absence of fresh meat we have excellent preserved beef of the first quality. If the supplies permit of it, I think, sir, that the ration of preserved meat might be augmented one-sixth. This measure, as well as that already taken, of supplying a daily supplementary ration of wine, seems to me useful, as long as we are subject to typhus, to strengthen the systems of the men against the attack of the epidemic.

"I do not speak of the powdered meat, which is considered very inferior to the preserved beef.

"I am surprised that no effort is made to use the flesh of the horses that are condemned to be killed. To overcome prejudice the example should be set by those in authority. A commencement might be made by distributing horse flesh to the officers. The generals might set the example by inviting the colonels to a dinner of horse flesh. This flesh is far preferable to the beef which we have here. If I did not fear that the use of it would affect the spirits of the patients, I would recommend it in the field hospitals.

"The spring will bring with it the invaluable dandelion. Meanwhile it would be well to encourage the excellent measures of the Minister of War for the shipment of large quantities of the fresh vegetables which abound at Constantinople.

"Work and moderate exercise are excellent aids to health. Nothing is more pernicious than absolute rest; idleness enervates the body and soul.

"Collections of rubbish have been allowed to accumulate around some of the camps; they should be carted away, as soon as possible, by well directed fatigue parties. The moment for redoubling care and cleanliness has arrived; the slightest violation of hygienic laws will increase the number of sick and deaths.

"For the time being, I confine my counsels to the regiments, to these few and simple measures.

"*Ambulances.*—You are aware, sir, that notwithstanding our united efforts, we have been unable to give our ambulances a good hospital status.

"Of the 6,000 mattresses put in use four months ago, there are hardly 2,500 which are serviceable, and it is with difficulty that the repairs are made in time to maintain this number.

"The barracks are only sufficient for 4,500 inmates.

"Blankets are numerous, but almost all of them contaminated; sheets are wanting, as well as means for thorough washing. Many of the sick are obliged to sleep in their trowsers. There are not sufficient camp beds for the tent hospitals, and many utensils are wanting, such as chambers, broth bowls, spittoons, etc., and above all the slippers and hospital clothing.

"I am not criticising; far from it. I am the first to acknowledge the great difficulties which have been overcome, before arriving at the present result. Besides, it must not be forgotten that we are carrying on a campaign in a country destitute of everything; I have only wished to prove the facts, and show that it is urgently required that the greatest number possible of our sick should be sent to Constantinople, in order to disencumber our field hospitals here. If it were necessary to demonstrate it by figures, I could find it in the afflicting results furnished by the sick treated in the Crimea, during the last few weeks.

"During the last ten days, from the 20th to the 29th of February, which is a good sample of the same period for a long time back, 519 left the ambulances cured, and 873 died!! A comparison in the cases of typhus gives a still more terrible result. For every 27 cures we have 383 deaths, and yet, sir, typhus in its *ordinary condition*, although a serious disease, does not carry off more than a sixth of the sick. For example, of 442 infirmary attendants attacked with typhus at Constantinople, only 42 died.

"These figures do not require comment; they loudly testify to the impotency of medicine, in the conditions found in the Crimea.

"Ought all the sick to be sent to Constantinople, or only a certain class? My advice is, to ship off all the non-typhic patients, if their condition allows of it; they are the most numerous. Their departure will effect an immediate disencumbrance of the hospitals and allow all the energies and care of the surgeons to be devoted to the unfortunate typhic patients, who, retained in the Crimea, will no longer risk the fleet and the hospitals of Constantinople with the dangers of infection.

"If the weather continues favorable, it would be advisable to place the typhic patients in tents, pitched at distances of 15 yards apart, and, if our resources will permit of it, to instal only two patients in each tent; they should not be allowed to lie on the ground, but be provided with camp beds, by this means they would be protected from the unhealthy influence of the lower stratum of the air; the bed clothing should be changed daily, and exposed to fumigating agents; whitewash the ground, fumigate, constant change of air, for I again repeat, contagious typhus must have pure air, constantly renewed, without which cure is impossible.

"You see, sir, military medicine is not very exacting, it can bend itself to the necessities of war; but the little we ask should be given without grudging.

"One of the grave considerations occupying M. Scrive and M: Mery, Medical Director of the 3d corps, is the sanitary condition of the two field hospitals, that of the 2d division of the 2d corps, in which the mortality at the present moment is very great, and that of the two first divisions of the reserve. I visited the last. Among the 600 sick, I found 319 cases of typhus in very deplorable conditions, so deplorable that, in my opinion, the immediate abandoning of the hospital can alone bring any remedy.

"Your Excellency can judge for yourself from the statement given below, handed to me by M. Goutt, Medical Director of this ambulance.

"*Field hospital of the first division of the reserve corps* (*3d corps*).

"Return of cases of typhus, from the 10th of January to the 10th of March, 1856.

Period.	Entered.	Left.	Sent away.	Died.
January 10 to 20	65	0	4	10
" 20 to 30	18	0	30	30
February 1 to 10	184	4	149	61
" 10 to 20	313	0	6	133
" 20 to 29	137	0	52	93
March 1 to 10	120	0	0	101
Total	837	4	241	428

"Of these 837 cases of typhus, 531 were sent directly from the regiments, and 306 were developed in the field hospital in the following proportion:

Chaplains	1	Sick who entered from other diseases	185
Administrative officers	0		
Attendants	85	Physicians	10
Laborers	25		

"The return of pleasant weather will, doubtless, modify in a favorable degree, this unfortunate condition, but the sun alone cannot accomplish all. We must come to its aid, and second its influence with our own efforts, if we wish to make sure of enjoying its benefits.

"The typhus of last year never presented the grave characters of the present attack. In the opinion of some physicians, it did not then even exist, but now, no one doubts its presence. The following total of deaths during the first six months of 1855, in the field hospitals of the Crimea, is highly instructive. It will be seen that the sun gave more activity to the emanations of organic miasms, and did not reduce the number of deaths, as we hoped it would.

	Deaths.	Effective Strength.		Deaths.	Effective Strength.
January	971	78,000	April	615	91,000
February	543	89,000	May	1,075	107,000
March	500	96,000	June	3,106	121,000

"The last figures embrace, it is true, a large number of deaths from cholera. In the above table, the deaths from wounds received in action, form but an insignificant part. In disclosing the sanitary condition of the army, loyally and as it actually exists, I feel that I have only done my duty. The situation is serious, but we can still control it, and your Excellency will surely triumph, if the simple measures proposed are adopted. The mortality is very considerable, amounting to five or six thousand men a month, in the Crimea and at Constantinople. It has hitherto mocked the wisest precautions; the burden of blame must be borne by the evils which war entails. The powerful assistance which your Excellency has rendered to the medical officers of the army, of whom nineteen have already died of typhus, and the high degree of confidence with which the profession has been honored and encouraged, have ever elevated its devotion and self-denial, and impressed me personally with the deepest gratitude."

X.

Report of Dr. Marcet, Chief of the Medical Service at Calchi.

NAVAL HOSPITAL AT CALCHI, *April* 10, 1856.

"INSPECTOR BAUDENS: SIR—I have the honor, in addressing you, to respond to your wish, that I would submit some notes upon the typhic affections that prevailed in the hospital at Calchi, from the latter part of December, 1855, to the 12th of April, 1856. You will please pardon the brevity of my statements, in view of the numerous engagements of the present time.

"I have had under treatment sailors attacked with the typhus in the following order:—

From the ship *Wagram*........................ 27	
" " " *Vauban*........................ 13	January
Sick attacked in the hospitals, particularly scorbutic patients.................................. 32	and Feb.
From the ship *Orénoque* 119	
" " " *Algeria* 129	
" " " *Magellan*, for the most part convalescent.................. 60	March.
" " " *Lucifer* 13	
Sick attacked in the hospital................... 53	

Total (of whom 64 died) ....446, or one death in seven.

"I attribute the very large proportion of sick from the *Algeria* and the *Orénoque* to the fact, that these ships were obliged to keep their sick ten or twelve days on board, during a cold and rainy period, which compelled all the passages for ventilation to be closed. It is worthy of remark, that the Algeria alone had any of its officers sick (five, one of whom was the commandant). The berths of the officers of sailing frigates are between decks, and have no air but what comes from the gun-deck, where the sick passengers lie. The commandant, whose berth was in the after part of the gun-deck, owed his sickness to his humanity. He had given up a part of his cabin to the passengers, and lived in the midst of the most severe cases. The masters of all the ships pay a large tribute to the disease being lodged in the lower parts of the vessel.

"The period of incubation of typhus appears to me to be from twelve to fifteen days. With few exceptions, typhus has presented three periods—premonitory, catarrhal, and that which I shall call the actual condition, during which the ataxic or the adynamic symptoms prevailed, sometimes both.

"The most general symptom was the petechial eruption, which was often of remarkable intensity. It showed itself during the first day, rarely after the fifth or seventh. Once it appeared on the ninth—the patient died two days afterwards.

"Pains in the head and epistaxis were the rule. The latter recurred five or six times in twenty-four hours; in three cases the excessive hæmorrhage necessitated plugging; they appeared to relieve the pains in the head. I observed bloody stools in only one case, and one case of hæmaturia, followed by very sedimentary urine and rapid improvement.

"Abdominal symptoms were slightly marked and rare (a distinctive character of typhoid fever); constipation very frequent, particularly at the commencement. The condition of the lungs deserves great attention in the catarrhal period; but, especially in the more advanced stages, I have noticed a hypostatic condition.

"As regards nervous symptoms, I have found them all, from the simple spasm to the most perfect form of epilepsy. I have seen one case of chorea, three or four contractions of the limbs, several of trismus, of convulsions of the muscles of the eye, and, finally, tonic convulsions of tetanus, or clonic action of eclampsia. Deafness has been frequently noticed. Delirium has always existed, four or five times of a furious kind; eight men were required to restrain the patient, in one case; the access lasted two hours, and appeared to succumb to inhalations of chloroform, and later to leeching behind the ears, and counter-irritants to the lower extremities.

"The progress of the disease is incomparably more serious, when strongly marked nervous phenomena exist, than in the worst stages of the adynamic condition.

"I have observed, as critical stages, abundant sweats; they did not appear to me to exercise a favorable influence;—parotitis (once double) almost always suppurated, good symptom, except in two cases, where the fever seemed to redouble in intensity, and caused death in twenty-four hours;—abundance of sedimentary urine, good symptom;—numerous abscesses on the body, boils, pustules of ecthyma, etc.; but of all these symptoms the sudamina have been the most favorable—their appearance has always presaged convalescence.

"More than fifty patients have had sores on the sacrum, or gangrenous spots on the limbs; in two cases these appeared on the scrotum. One of the patients, who seemed to be convalescent, was affected with sphacelated ulcers of the scrotum, of the upper part of the thigh, and of the perineum, which eat through into the rectum in two places; he died. Once I observed rapid ulceration of the two corneas.

"*Treatment.*—At the beginning, an emetic or emeto-cathartic, with an infusion of flax-seed. During the catarrhal period, diet, sudorific and diluent drinks, a few leeches to the mastoid process; saline or oily purgatives. When the adynamic stage is reached, broth, drink of wine, tisane, with melissa or acetate of ammonia, eight to thirty grammes in twenty-four hours; frictions with camphorated vinegar; to stop involuntary stools, enemas of hypochlorate of soda, or ratanhia, which is preferable; if there is distension of the stomach, I employ camphorated embrocations. When nervous symptoms predominate, I have employed camphor, musk, ether, but not often with any good result.

"At the commencement, I bled twelve or thirteen men—only one died. In three cases the blood was covered with a thick buffy coat."

This report is particularly worthy of attention, because the sailors who were attacked by the disease were in perfect health, and of unimpaired constitution. Likewise, the progress of typhus has been more distinct and regular in the crews of the navy than in the worn-out soldiers; and the blood-letting, which would have been fatal to the latter, was employed with benefit to the former. Bleeding at the nose has been more frequent than with the soldiers. We have likewise observed gangrenous eschar, following typhus, in the hospitals of the land forces. As to the furious delirium which M. Marcet had such trouble in suppressing, we have seen Medical-director Garrat, at Daoud-Pacha, constantly overcome it with ease by a combination of opium and sulphate of quinine, administered as a drink.

XI.

Report to the Minister of War.

CONSTANTINOPLE, *April* 28, 1856.

" MARSHAL:

"The excellent manner in which our hospital service is carried on, reveals itself in results more and more satisfactory. The mortality decreases notably, and at the same time the cures are on the increase. Medicine is not alone in congratulating itself upon a condition so advantageous to the success of its therapeutics. Surgery also is benefited thereby. Amputations, especially those practised for frost-bite, were almost always followed by hospital gangrene, and ended in death; now they almost all succeed.

"Some unfortunate fellows whose feet had fallen off from the effects of frost-bite, have suffered amputation of both legs, with perfect success. Our hospital establishments, to which considerable additions have been made, feel no longer the influence, so to speak, of epidemics or miasms; the new hospitals are nevertheless the most favored in this respect. Thus, in the large barrack hospital of Ramis-Tchifflick there is comparatively no mortality, and no cases of typhus.

"The large field hospital in tents, established in a few days at Prinkipo, is already full of scorbutic patients, whose faces gleam with joy at finding themselves in such healthy quarters.

"Scurvy has been the precursor of typhus: it was urgent to withdraw from the influence of this disease the scorbutic patients on whom it throve.

"Four cases of typhus appeared among the scorbutic patients; the seeds of the disease were sown in our hospitals, and the incubation of the infectious miasms took place after their arrival. I caused a few tents to be pitched at a distance of a quarter of a mile, and there placed the typhic patients. By these means we shall certainly escape its propagation to the other sick.

"I estimate that in twenty or thirty days the 1,800 scorbutic patients now at Prinkipo will be sent off, and their places occupied by others. When the fresh arrivals of scorbutic patients cease, we can begin to receive convalescents.

"I intend to subject the convalescents to a quarantine of fifteen days in another part of the island, and shall only embark them for France, when I am convinced after this lapse of time, they are insured against an attack of the reigning epidemic.

"Our last report, up to the 27th instant, of the typhus at Constantinople, which is just handed to me, presents the following figures:

	In Hospital.	Entered.	Cured.	Dead.
April 27	919	29	113	11

"This result is very satisfactory; the comparison of the number of cures with the deaths fully confirms the value of the precautionary measures before mentioned.

"I think, Sir, that the subject which should occupy our attention at the present time, is the manner of transporting the troops from the Crimea to France.

"In a conference with General Espinasse* it was decided, in order to equalize the distances between the Hospitals on the route followed by the fleet, that field hospitals in tents should be established at the Pirea and at Messina. Difficulties are placed in the way by the authorities at the latter place, who are opposed to a depôt for typhic patients being established there. We shall report on the Pirea and on some island as near as possible to Messina.

"I may be obliged to visit Greece, to find a convenient depôt. The space assigned to the present quarantine at the Pirea is altogether insufficient—forty

* The Emperor had charged General Espinasse, on his leaving the Crimea to return to France, with the mission to inspect the Hospitals at Constantinople.

typhic patients would certainly engender infection. Dr. Artigues, who has resided for some time in this country, has given me valuable information, which I shall make use of. And although I hold a sea voyage in great horror, I should not hesitate to make it, in the hopes that some service might result from it.

"I am at this moment much distressed. My private secretary, M. Benjamin Crombez, a young man full of courage and self-denial, whose help is so valuable to me, is suffering from a severe attack of typhus, caught during our visits to the hospitals. I pray heaven to take pity on him and spare me the pain of seeing him die a victim to his devotion!

"I have this moment received from Marshal Pelissier an answer to my letter of the 22d; he tells me that the reports received from all quarters agree in showing that the sanitary conditions are considerably improved. He continues to use all efforts to advance this improvement, for which happily the weather is favorable.

"I have always found the Marshal ready to listen to the counsels of the physicians; he gives them full justice and honors them with his fullest confidence. I owe to his assistance the prompt execution of the prescriptions which I deem indispensable.

"I learn from M. Scrive that all the prophylactic measures applied to the troops before their departure for France, are faithfully executed. He is strongly impressed with the danger of propagation of typhus on board of the transports on their passage to France.

"In one of his letters, he says: 'I yesterday received a report from Mr. Molard, director of the hospital at Gallipolis; sixty-two cases of typhus have been landed from the ships *le Navarin* and *le Jupiter*, and yet the soldiers left Kamiesch and Eupatoria in good health. It is impossible that it should be otherwise. What would have been the result if these vessels had remained fifteen days at sea without touching land?'

"From General de Martimprey, chief of the staff, I received a letter on the 2d of May, 1856, as follows:

"'You know, sir, that we are improving here, although slowly. Those remaining of the 2d corps, which is now embarking, are in little shelter tents, on ground which has not before been occupied, and remote from the other encampments. The guard and the 1st corps, which are to embark after, are subjected to the same precautionary measures. On receipt of your letter of the 22d of April, the most stringent orders were again given for the punctual enforcement of the hygienic measures which you advise. You would find that the division of the reserve corps, at this moment suffering from typhic infection, is encamped on the sides of the high mountain which overlooks the plain of Balaclava.'

"In conclusion, I am happy to be able to assure your Excellency, that the epidemic is on the decrease, and that the army has nearly reached the end of its terrible trials."

XII.

The necessity of breathing a pure and continually renewed air.

To prove how necessary pure air is to man, as well as to all organized beings, we shall confine ourselves to a few facts borrowed from some articles by M. Boudin, in the "Annales d'Hygiene publique et de Medicine legale":
"In 1834 a spacious and handsome house was constructed in the Zoölogical Gardens, London, intended for the monkeys, about 30 in number, who for several years past had lived in pretty good health, in the open air. The principal object in constructing this new building was to provide them in winter with an artificial warmth similar to that of their native clime. A few weeks after their establishment in the warmed building two-thirds of them died by the disease which kills the majority of our soldiers, pulmonary phthisis; the survivors were at the point of death. *It had been omitted to renew the air by ventilation.* Fresh air was liberally supplied, and the monkeys which had not yet succumbed, soon recovered."

The same author cites the following extraordinary occurrence: "The Hospital Beaujon in Paris is composed of four separate buildings, of exactly the same size, and each containing the same number of patients, suffering from analogous complaints. During several years hospital gangrene, erysipelas, and pleurisy were rampant in three of the buildings; one building only was exempt, it was the one ventilated by the Leon Duvoir system."

The Council of Health of the Army was consulted fifteen years ago, on account of a murderous attack of typhoid fever, which occurred yearly at the St. Cloud barracks, as soon as King Louis Philippe took up his residence at that place. None of the inhabitants of the town or the officers were affected, and as soon as the king left the place the epidemic ceased. It was accounted for in this manner: while the king was away the barracks contained 400 men only; but when he arrived, the number in the same badly ventilated barracks was increased to 1,200.

We hasten to add that the War Department, profiting by the severe lessons of the past, has given strict attention to the ventilation of hospitals. In the splendid military hospital which has just been built at Vincennes, the apparatus for heating and ventilation are so arranged that each patient is assured a minimum of *sixty cubic yards of air per hour*, and a uniform temperature, day and night, of 60° Fahr.

These improvements, so creditable to the Minister who has thus taken the initiative, cannot fail to secure the best results, and we sincerely hope that the same principles will, before long, be applied to all the barracks.

THE END.

INDEX.

Acclimatization, 63.
Acids, vegetable, 36.
Acrodynia, 149.
Air, poisoned in tents, 169.
" in closed cavities, 88.
Algeria, surgery in, 215.
" hospital gangrene, land, 13.
" ship, 245.
Ambulances, 241.
" English, 111.
" French, 112.
" Russian, 111.
" not as good as chairs, 135.
" of the trenches, 62, 64, 66.
American bread rations, 30.
" beef, 33.
" desiccated veg., 35.
Amputations at joints, 90.
" experiments deprecated, 89.
" faulty, 89, 248.
" former practices, 89.
" immediate or delayed, 90, 92.
" low as possible, 89.
" partial, 99.
" rules for, 88, 99.
" spontaneous, 98, 100, 224, 225, 226, 227.
" statistics of, 213, 214, 215.
" sub-malleolar, 89.
" thigh, 92.
Antiscorbutics, 35, 36, 39, 60, 180, 189.
Apothecaries, French army, 66.
Arabas, 26, 51, 124, 127, 150.
Arms of precision, 78.

Army of Potomac, sickness in, 142.
Aromatic fumigations, 170, 172.
Attendants from convalescents, 176.

Baidar, 14, 17, 19, 20, 25, 26, 32, 33, 46, 56, 163.
Bailly, Dr., 124.
Bakeries, camp, 30, 56.
Balaclava, 14, 15, 16, 17, 20, 21, 35, 73, 76, 192, 196, 251.
Balls, conical, 78, 79.
" in wounds, 82.
" (see Gunshot Wounds.)
Bandages, 68, 78.
Barracks, 28, 50, 58, 169.
Baths of Agamemnon, 11.
Battle, Alma, 135.
" Balaclava, 20.
" Inkermann, 16, 18, 19.
" Traktir, 18, 19.
Baudens', M., fracture box, 91, 92, 215–223.
" report to Minister, 248.
Beaujon hospital, 252.
Beds and bedding, 49, 50, 59, 68, 172, 185, 241.
Beef, 31, 33, 41, 110, 205.
Beer, 75.
Birds of prey, 65, 125.
Biscuit, 29, 30, 59.
Bivouac before voyage, 192.
Blankets, 52, 59, 68, 185, 241.
Bolting of flour, 31.
Bombs, wounds by, 87.
Bones, use of, 31, 32.
Boots, 23, 54, 55, 141.
Bowel complaints, English, 140,

Brandy, 37, 38, 59, 75.
Bread, 27, 30, 31, 40, 59, 75, 115.
Breakfast for troops, 40.
Bronchitis, 51, 53, 151.
Building materials, 50, 51, 58.
Burden of a soldier, 55.

Cabbage, 34, 205.
Cacolets, 135.
Calchi, hosp. island, 200, 244, 247.
Callosity of fractures, 92.
Camels for carrying wounded, 135.
Camp, 25, 43.
" of 81st Line, 49.
" change of, 44, 57, 185, 192, 198.
" fires, 46, 47.
" Russian, 47, 107.
Castle hospital, 73.
Catarrh, 47.
Cautery in hospital gangrene, 95, 96.
Cavalry cantonment, 21.
Cemetery, army, 44, 65.
" Sebastopol, 22.
" Turkish, 119.
Champaigne, 75.
Chapel, head-quarters, 26.
" hospital, proposed, 65.
Chicory, 110, 205.
Chlorine, 44, 63, 170, 172, 173, 181, 185, 190, 203.
Chloroform, 104, 105, 106.
Chocolate, 68.
Cholera, 10, 11, 27, 43, 121, 123, 124, 125, 126, 127, 128, 129, 145, 180, 244.
" English loss by, 197.
" propagation of, 131.
" symptoms, 134.
" theories, 129.
" treatment, 130, 131.
" unlike typhus in attack, 175.
Cleanliness, 123.
" English, 45.
Climate of Crimea, 25.
Clocheton hosp., 64.
Clothing, 51, 58.

Coffee, 38, 59.
Cold, action of, in inflammation, 84, 85.
Communications, friendly, 17, 18, 23, 195.
Compresses, 68, 78.
Condiments, 60.
Conservative work of nature, 99.
Constantinople camps, 43, 162.
" citizens escape typhus, 201.
" convalescent depot, 164.
" hospitals, 28, 32, 43, 45, 71, 72, 107, 118, 119, 131, 134, 138, 170, 174, 203, 204, 205, 213, 214, 218, 227, 228, 235, 249.
" resources at, 28, 29, 35, 171.
" sanitary condition, 56, 164, 180, 188, 228.
" school of medicine, 111, 202.
" scurvy in, 199.
" sick sent to, 28, 63, 112, 115, 119, 132, 137, 144, 161, 162, 241, 242.
" surgery, 90, 91, 100.
" typhus, 165, 166, 167, 168, 172, 176, 178, 181, 182, 183, 184, 188, 189, 190, 191, 199, 201, 242, 243, 244.
Contusions, treated with ice, 86.
Convalescent depot, 28, 164.
" attendants, 176, 178.
" camps, 234, 237.
Cooking, 35, 36, 37.
Cotton for dressings, 68, 210.
Council of health, 187, 210, 252.
" on typhus, 172.
Councils, medical, 101, 102.
Criméenne, garment, 51, 52, 53, 58.
Crowding, how understood, 165, 166.

Danube, 124, 125, 129.
" army of the, in 1829, 76.

Daoud-Pacha, 118, 137, 138, 236, 237, 248.
Depot, Maslak, rules, 137, 234.
Descartes, steamer, 14.
Desiccated vegetables, 34, 35.
Diarrhœa, 33, 53, 58, 82, 121, 180, 138, 139, 140, 150, 228, 231.
Diet drinks, 139.
" hospital, 205, 211.
Digestion, theory of, 34.
Dilatation of conical balls, 79, 80.
Disarticulation, 90, 92.
Discipline promotes health, 189.
Disinfectants, 44, 63, 173, 181.
Dislocations treated with ice, 86.
Distortions cured by use of apparatus, 218, 219, 220, 221, 222, 223.
Dobrutcha, 121, 122, 123, 124, 125, 127, 129, 131, 143.
Dolma Batché, 136, 236.
Drainage of camps, 49.
Dramatic entertainments, 195, 196.
Dumas on digestion, 34.
Duvoir, Leon, ventilation, 252.
Dysentery, 27, 38, 53, 139, 172, 197, 228.

Economy often pernicious, 42.
Edinburgh, typhus at, 155.
Egyptian troops, 109.
Electric wires, 24, 26.
Embarkation of troops, 134, 199.
English careful of soldiers, 196.
" losses by war, 197.
" sickness, bulletin of, 198.
Epidemics of the Crimea, 113.
Erysipelas, 252.
Eupatoria, 15, 108, 109, 121, 147, 151, 175, 233, 250.
Expenses of a company, monthly, 42.
Explosion of magazines, 69.
Extemporised dressings, 103.
Extraction of balls, 83.

Fatalism concerning typhus, 192.
Felchers, 77, 78.
Fêtes, military, 195.

Fevers, 228.
" Crimean, 141, 198.
" intermittent, 13, 121, 141, 150.
" remittent, 13, 141, 142, 150.
" typhoid, 40, 41, 46, 56, 170, 172, 197, 229, 252.
" typhus (see Typhus).
Field of the dead, 143.
Firemen, 41.
Fish, 33, 153.
Flannel girdle, 53, 58, 131, 133, 139.
Food, 29.
Fortifications built, 114.
Fowls, 33, 76.
Fractures, apparatus for, 90, 91, " 92, 216, 217, 218, 219, 220, 221, 222, 223.
" ice, in treatment of, 86.
Franka, 120, 124, 127.
Freezing, 96, 97, 98, 224, 225, 226, 227, 228.
Fuel, 25, 26, 27, 30, 46, 56, 57, 133.
Fumigations, 110, 160, 170, 172, 181, 203, 239.

Gaiters, 59, 60.
Gallipolis, 12, 113, 114, 115, 116, 118, 123, 174, 206, 250.
Game, 33.
Gardens, culinary, 36, 60, 62.
Gelatine of bones, 32.
Geology of Crimea, 14, 18, 20.
Grain raised by troops, 50.
Greaves, 54.
Grottoes as hospitals, 65.
Gulhané, 172, 173, 203, 215, 216, 217, 224, 225, 236.
Gunshot wounds, 80, 81, 82, 83, 84, 85, 93, 94, 215, 216, 217, 218, 219, 220, 221, 222.

Habit, Russian, 52.
Harem, hospital of, 205.
Helmet, fur, 53.
Hæmorrhage checked in disrupture, 87.
Hernias treated with ice, 86.

Hildebrand on typhus, 157, 158, 161, 184.
Horses, Arabian, 195.
Horse-flesh, 32, 33, 240.
Hospital, Adrianople, 116.
" artillery park, 62.
" arrangement, 63, 64.
" attendants, 186.
" Bachistach, 205.
" Balaclava, 73.
" Balbec, 48.
" barracks for, 67, 70.
" Beaujon, 252.
" Caulidjé, 136.
" Castle, 73.
" Calchi, 200.
" Constantinople (see Constantinople).
" convalescent, 106.
" Daoud-Pacha, 118, 137, 138, 168, 236, 237, 247.
" Dardanelles, 12.
" diet, 68, 74, 110.
" Sardinian, 211.
" Dolma Baktché, 136, 236.
" division, 65, 66, 68.
" English, 11, 12, 70, 73, 74, 75, 196.
" Eupatoria, 109, 110.
" field, 27, 28, 62, 63, 64, 65, 66, 67, 69, 126, 127, 135.
" frigate, 196.
" Gallipolis, 12, 115, 174, 250.
" Gulhané, 90, 137, 203, 215, 216, 224, 225, 236.
" Guyer, 194, 206.
" harem, 205.
" head-quarters, 26.
" Hyeres, 194, 206.
" injured by explosion, 70
" Jeni-Batché, 203.
" Kamiesch, 22.
" Malta, 10.
" Maltepé, 118, 119, 236, 237.
" Marseilles, 14.
" Maslak, 177, 180, 181, 187.
" military school, 138, 236, 237.
" movements of, 68, 69, 70.
" Navy office, 203, 205.

Hospital, overcrowding perilous, 108.
" Parade-ground, 138, 219, 220, 221, 222, 223, 236, 237.
" Pera, 143, 144, 236.
" Piedmontese, 76.
" Pirea, 206.
" Prinkipo, 199, 200, 248.
" purification of, 177, 180.
" Ramis-Tchefflick, 137, 236, 237, 248.
" register, 64.
" Rodosto, 118.
" Russian embassy, 76, 111, 137, 145.
" Saint George, 74.
" St. Marguerite, 199.
" Sardinian, 74, 75.
" Séraskiérat, 203.
" Sisters of Charity, 145.
" series of, to France, 206, 249.
" Smyrna, 11.
" supplies to, 68.
" tents for, 48, 67, 70, 127, 128.
" trenches, of the, 64, 65, 66, 69.
" Turkish, 109, 110, 147, 203, 204, 205.
" University, 137, 236.
" Val-de-Grace, 40.
" Varna, 121.
" Vincennes, 252.
" Visited, 174.
" washing for, 68.
" gangrene, 12, 13, 27, 43, 86, 94, 100, 145, 158, 248, 252.
" described, 94, 95.
" isolation required, 95.
" treatment, of, 95, 96.
Hungary fever, 155.
Hunter's experiments, 85.
Huts, 46, 56, 58, 73, 74.
Hygiene, ignorance of, 12, 43.
" precautions in, 60, 61, 64, 193.
" statistics of, U. S., 61.

Ice on gunshot wounds, 84, 85, 86.
" theory of action of, 85.
Imperial Medical Society, 202.
Indians, survive greater injuries, 144.
Infirmaries, 62, 63, 64, 67.
Inflammation after wounds, 89.
Infusions, medical, 139.
Inkermann, 16, 17, 61, 73.
Inspection, 14, 26, 66, 111.
Instructions in case of epidemics, 167.
Jupiter, ship, 250.

Kamiesch, 14, 15, 16, 21, 22, 23, 33, 57, 192, 196, 198, 250.
Kearney's brigade bakery, 30.
Kidney beans, 36.
Knapsacks, 55.
Kustendjé, 124, 125, 126.

Labor promotes health, 41, 42.
Laborers, English, 41.
" French, habits of, 41.
" in Algeria, 42.
Land, 32, 59.
Larrey, writings of, 103.
Lazarettos as hospitals, 10, 12.
Lemons, 60, 153.
" juice, 36, 39, 189.
Lentils, 36.
Letters, M. Angot to Baudens, 164.
" Baudens to Marshal Pelissier, 168, 177, 189, 207.
" to Minister of War, 27, 55, 174, 181, 193.
" Council of Health to Baudens, 187.
" Intendant to Baudens, 186.
Liebig on digestion, 34.
Ligature of arteries, statistics, 214.
Lime water, 44, 62, 185.
Linen for dressings, 68, 209, 210.
Lisfranc on wounds, 82.
Macadamized roads, 25.
Macleod, Dr., 90, 106, 140, 214.
Magendie on varied food, 40.
Mahommedan indolence, 114.

Malarious influences, 142, 143.
Maltepé, 119, 171, 236, 237.
Mangalia, 123, 125, 126, 127.
Marcet, Dr., report, 244, 247.
Marching orders, Gen. Bugeaud, 132.
Marseilles, 14, 28, 99, 179, 183, 184, 186, 194, 201, 205.
Mascara expedition, 38, 135, 215.
Maslak, 119, 138, 145, 161, 171, 180, 181, 187, 228, 229, 235, 236, 237.
Mayence, fever, 155, 161; siege, 175.
Meals, number should be increased, 39.
Meat, horse, 32, 33, 240.
" powdered, 32, 240.
" preserved, 32, 59, 240.
Medical education, Turkish, 201.
" " Russian, 77.
" society founded, 202.
Messina, 206, 249.
Mexican war statistics, 61.
Miasms, 15, 43, 44.
" fevers from, 141.
Military school hospital, 138, 236, 237.
Milk concentrated, 27, 68.
" fresh, on ship board, 196.
Mines and minerals, 153.
Minié balls, 78.
Morale of troops, 208.
Mutton, 32, 110, 205.
Mytilene, 151, 152, 153, 154, 171, 200.

Naples fever, 155.
Navarin, ship, 250.
Negligence, French soldiers, 45, 63.
Nervous prostration, 104.
Nightingale, Miss, 74.
Nitrogenous foods, 34.
Nostalgia, 28.
Nurses, female, 76, 77.

Offal, disposal of, 50.
Onions, 60.
Orénoque, ship, 245.

258   INDEX.

Parade-ground, hosps., 138, 173, 219, 220, 221, 222, 223, 236, 237.
Paris, bakeries, 30.
" medical schools, 111, 147.
Patriotism, Russian, 150.
Pera, 137, 138, 143, 145, 146, 173, 183, 236.
Percy, quoted, 103, 184.
Perforations of foot, by balls, 94.
Periosteum generates bone, 93.
Peritonitis, 228.
Peyer's glands, diseased in typhoid, 232.
Pharmaceutical department, 138.
Philadelphia, typhus at, 155.
Phthisis pulmonalis, 41, 205.
Physicians (see Surgeons).
Piedmontese, 19, 20, 46, 76, 211.
Pirates, haunts of, 20.
Pits for huts, 14, 15, 46, 47, 56, 57.
Plague, 1792, 174.
"   1829, 76.
Pleurisy, 262.
Pneumonia, 53, 150.
Police, 25, 146.
Policnity, 153.
Poucharra arms, 79.
Port-Vendres, sick at, 179.
Posen, typhus at, 155.
Potatoes, 33, 50, 60 75, 153.
Preparations for sickness, 168.
Preserved vegetables, 27, 33, 34.
Prinkipo, Island, 187, 199, 200, 248, 249.
Proximate principles, 34.
Puddings, 74.

Quarantine, 194, 200, 205, 249.
Quarries, 23.
Quindros thermal spring, 152, 153.
Quinine, use of, 142.

Radishes, 60.
Ramis-Tchifflick, 118, 171, 174, 236, 237, 248.
Rations, hospital, 211.
" English, 39.
" French, 29, 39, 41, 145.

Records, statistical, 164.
Regimental fund, 30.
Resection, 92, 93, 94, 213, 214, 215.
Resources of the Crimea, 25.
Return of army, 195, 205, 249.
Rewards for preserving horses, 37, 50, 59, 68.
Rheumatism, 46, 47.
Rice, 205.
Rifled arms, 79, 80.
Rifles, Russian, 79.
Rum, 75.
Russian embassy, 145.
"   prisoners, vitality of, 144.

Sac formed around balls, 83.
St. Cloud barracks, 252.
St. George, monastery, 14, 21, 28, 73, 198.
St. Omer, sickness at, 37, 41.
Sanitary condition, 56, 189.
Sardinian hospitals, 75, 211.
"   troops, 20, 195.
Sausage, 32.
Scarcity produces sickness, 40.
Scenery, Sebastopol, 106.
School of Medicine, Turkish, 201.
Scorbutic taint, 140, 150, 170.
Scrive, M., 71, 102, 106, 161, 162, 163, 179, 199, 242.
Scultet, apparatus of, 222, 223.
Scurvy, 13, 20, 27, 35, 40, 41, 47, 50, 56, 60, 76, 109, 146, 148, 149, 158, 180, 199, 200, 228, 248.
"   causes, 148.
"   epidemic, 150,
"   English army, 150.
"   Russian army exempt, 150.
"   symptoms, 148.
"   treatment, 150, 151.
"   Turkish army, 151.
Scutari, 80, 116, 143, 203, 204.
Sheepskin garments, 53, 54.
Shells, wounds by, 86, 88.
Shelters, 56.
Shelter-tents, 45, 47, 59, 135, 198, 239, 251.
Shirts, woollen, 53, 54, 59.

INDEX. 259

Shoes, 54, 133.
Sicily, hospital prohibited, 206.
Sickness, aggregate French, 207.
Sidi-Ferruck, wounds at, 81.
Silistria, 115, 122, 129.
Sisters of Charity, 10, 76, 145, 149, 186, 229.
Sisterhood of St. Vincent de Paul, 183.
Skeletons of the slain, 17.
Soil of Crimea, 14, 15.
Soldier-dressers, 71, 72, 76.
Soldiers, careless of health, 121.
Soup, 40, 58, 60.
Sour-krout, 36, 60.
Spencer, English, 53.
Splinters of bone, 82, 92, 216, 217. 218, 219, 222.
Staoueli, gunshot wounds at, 81.
Stockings, woollen, 54, 59, 60.
Strasbourg, typhus at, 155.
Students, Turkish, 201.
Surgeons, allowances to, 187.
" assistants, 72, 73, 76, 77, 110, 186.
" badges of, 18, 73.
" devotion of, 9, 71, 100, 101, 171, 176, 207.
" died of typhus, 182, 233.
" discussions among, 101, 102.
" division hospitals, 66, 69, 70.
" duties of various grades, 67.
" English, 73, 74, 106
" funerals of, 183.
" grades, French, 71, 72, 77.
" losses of, 179, 182, 188, 193.
" military, 66, 67, 186.
" pensions not allowed, 183.
" promptness necessary, 103.
" qualifications of, 102, 103.
" quartered in hospitals, 70, 71.
" Russian, 77, 78.
" Sardinian, 106.
" Turkish, 201, 202.

Surgeons, wounded, 18.
Sultan, 116, 118, 143, 146, 147, 148.
" attends a ball, 146.
" endows a medical society, 202.
Surgical operations, 78.

Tchernaïa, 14, 16, 17, 18, 19, 20, 25, 44, 69, 163, 195, 198.
Tea, 11, 38, 74.
Tenedos, associations of, 12.
Tents, conical, 59, 147.
" Crimean, 47.
" hospital, 48.
" marquee, or wall, 48.
" shelter, or "Tente-abris."— (see Shelter-Tent.)
" ventilation of, 44.
" winter use bad, 169.
Terrassacum, plant, 35.
Thiéry, method of washing, 136.
Thouvenin arms, 79.
Tige rifle, 79, 80.
Tlemcen, expedition of, 38.
Tolerance of Turks, 138.
Topography of Crimea, 10, 14, 20.
Torgau, typhus of, 161.
Tornado, 15.
Toulon, sick at, 155, 179, 183, 186, 193, 201, 205.
Trajan, camp of, 124.
Tunic, 52, 53.
Turbot, buckle, 83.
Turks, ablutions of, 45.
Typhic condition, 229.
Typhus, 15, 20, 27, 35, 40, 43, 47, 71, 74, 76, 138, 140, 145, 155, 174, 248, 249.
" alarm in south of France, 205.
" Calchi, 245.
" Constantinople, 229, 230, 231, 232, 233.
" compared with cholera, 175.
" convalescence, 159, 232.
" course indefinite, 157.
" decline of, 207.
" differs from typhoid fever, 232, 233.

Typhus, disencumberment, 177, 180, 181, 187, 190, 201.
" English, 196, 198.
" history and origin, 155, 156, 238.
" limited to hospitals, 179.
" Marseilles, 194.
" not to be sent off, 184, 242.
" nurses die of, 183.
" officers escape, 171.
" premonitions of, 158, 238, 246.
" prevention, 167, 172, 174, 238.
" progress of, 157, 158, 159, 165, 166, 168, 171, 177, 181, 184, 186, 189, 190, 191, 199, 242.
" removal of causes, 192.
" reports on, 177, 237, 243, 244, 245.
" should be kept in the Crimea, 177, 186, 188.
" sent to France, 183.
" sent to camp, 182.
" ship-board, 184, 192, 194, 200, 201, 246, 247, 250.
" surgeons die of, 171, 182, 183.
" symptoms, 155–158, 230, 231, 238, 246, 247.
" synonyms, 155.
" treatment of, 160, 191, 247.

Undeveloped disease, 140.
University hospital, 236.
Utah, troops in, 35, 48.

Val-de-Grace, 86, 103, 124.
Varna, 26, 57, 109, 116, 117, 118, 120, 122, 124, 125, 126, 127, 128, 151, 180.
Vegetables, fresh, 35, 50, 60, 241.
Ventilation, necessity proved, 251.
" of camps, 44, 47, 49, 56, 57, 160, 185, 190, 239.
Vincennes mil. hosp., 252.

Washing, 68, 138, 173, 204, 209, 210.
Wind of Crimea, 15, 25, 44.
" of a ball (so called), 88.
Wine, 11, 37, 74, 115, 139, 153, 181, 212.
Winter quarters, 26, 29.
Wooden shoes, 54, 59.
Woollen under clothing, 53, 54, 59.
Wounds, battle of Alma, 135.
"   " Inkermann, 16.
"   " Traktir, 19.
" siege, 24, 65, 66, 68, 69.
" shell, 80, 81, 86.
" gunshot, 80.
Wringing machine, 204.

Zouaves, costume of, 52, 118.

www.ingramcontent.com/pod-product-compliance
Lightning Source LLC
Chambersburg PA
CBHW021345230426
43666CB00006B/412